Gravitational Waves

A History of Discovery

Gravitational Waves

A History of Discovery

Hartmut Grote

CRC Press
Taylor & Francis Group
Boca Raton London New York

CRC Press is an imprint of the
Taylor & Francis Group, an **informa** business

Originally published as Gravitationswellen: Geschichte einer Jahrhundert-entdeckung (Beck'sche Reihe 2879) (German Edition) by Hartmut Grote © Verlag C.H.Beck oHG, München 2018. This is an updated and extended version.

CRC Press
Taylor & Francis Group
6000 Broken Sound Parkway NW, Suite 300
Boca Raton, FL 33487-2742

International Standard Book Number-13: 978-0-367-13681-9 (Paperback)
978-0-367-13682-6 (Hardback)

Visit the Taylor & Francis Web site at
http://www.taylorandfrancis.com

and the CRC Press Web site at
http://www.crcpress.com

Cable speaks

I'm just a cable, I kiss your feet
but I require a little deed
just give me strain relief and name
I shall return to you the fame
through signals passing undisturbed
of rippled space-time mightly curved

Contents

Preface

The first detection of gravitational waves on September 14, 2015 significantly expanded the scientific sensorium and allowed for the wondrous activity of the universe to be 'audible' for the first time. This event, the result of the highly energetic fusion of two black holes, is epochal for physics and, in recognition of their contributions to this accomplishment, the Nobel Prize 2017 was awarded to Rainer Weiss, Kip Thorne, and Barry Barish.

The signal detected on earth was sent on its way more than a billion years ago by two black holes of about 30 times the mass of our sun. During the furious act of merging to form a single black hole, this event released more energy than all of the stars in the entire visible universe combined. This energy would travel at the speed of light as a distortion of the fabric of space-time: a gravitational wave. When arriving on earth the stretching of space-time was no more than one part in a thousand billion billion and lasting for less than a tenth of a second. Yet, after decades of work of hundreds of scientists and engineers, the technology was just ready to catch this feeble whisper. We now know that, in a broad sense, events similar to this one happen about every 5 minutes throughout the observable universe, which illustrates how much this new field of astronomy is just at the start of an exciting journey.

The goal of this book is to convey the history that led to this first detection of gravitational waves in a way that is understandable to nonspecialist readers. As the story unfolds, some aspects of modern scientific research will also be revealed. The detection of gravitational waves illustrates the tools currently used by astronomers and astrophysicists such as large measuring equipment, elaborate computer simulations, sophisticated statistical analysis methods, and international research collaborations. Discoveries are seldom made in a continuous or linear fashion, but rather progress in spurts and often take detours. Programmatic decisions may stem from individual choices, which are technical or political in nature, yet can have a profound impact on future discoveries as will be seen in this book.

Chapter 1 begins with Newton's law of gravity and unfolds into Einstein's theory of relativity. The general theory of relativity could explain a long-standing anomaly in the planetary orbit of Mercury, and it predicted the bending of light by the sun, an effect that was spectacularly confirmed by observations in 1919. The prediction of gravitational waves from Einstein's theory proved to be quite difficult and controversial though. For many years, it remained an open problem as to whether or not these waves even exist or could ever be measured. The chapter concludes with a brief overview of the astronomical objects that are known or expected to generate gravitational waves.

Chapter 2 provides a historical outline of the first attempt by Joseph Weber to measure gravitational waves in the 1960s. Weber's assertion that he had actually measured gravitational waves inspired researchers around the world to attempt to reproduce his results, yet to no avail. Although the prevailing view was that Weber must have made mistakes, the detectors he started developing were refined and utilized over decades, and Weber can truly be regarded as the pioneer of gravitational-wave detection.

Chapter 3 describes the emergence of interferometer technology in the 1970s as a possibly more promising way of measuring gravitational waves. Over time, several improvements to a type of interferometer named after the physicist Albert Michelson increased the sensitivity of these instruments enormously and successively larger instruments were built. The most important prototypes of these interferometric devices were developed in Germany, Great Britain, and the United States.

Chapter 4 gives a compact overview of the history, working principles, and features of the large interferometers around the world. Systems with three to four kilometer-long evacuated beam tubes, or arms, were developed in the 1990s in the United States and in Italy; smaller facilities were built in Germany (600 meters) and Japan (300 meters). In Japan, an interferometer with three-kilometer arms is currently under construction (2019) and is expected to start operations soon.

Chapter 5 explores the question of how to actually look for signs of gravitational waves in the data provided by the detectors. Depending on the astronomical sources (such as neutron stars or black holes), specific waveforms can be expected that are calculated with complex computer simulations. The varied patterns, or signatures, of the respective waves are subject to a clever analysis procedure (blind analysis) in which all parameters are defined before the actual analysis happens. As a different form of blind analysis, fake artificial signals, called blind injections, may be intentionally added to the data stream. These procedures can help ensure that any measurement and analysis results are not distorted by the expectations or assumptions of the scientists involved.

Chapter 6 describes the historic moment, on September 14, 2015, of the first direct measurement of a gravitational wave hitting the earth. It was registered by two detectors in the United States. This event sparked feverish activity within the international scientific collaboration, continuing for more than five months, ranging from addressing whether or not it could have been a mistake or, possibly, a prank, to wrangling over the final title of the journal article. By the end of 2017, signals from 5 pairs of coalescing black holes had been published, and in 2018 that number was increased to 10. A new field of physics, gravitational-wave astronomy, was born!

Chapter 7 addresses research questions of gravitational-wave astronomy and presents an overview of some of the challenges and possible solutions to make the detectors more sensitive. In addition to improvements to existing equipment, new detectors must be built in order to explore the field more deeply. Along with new terrestrial detectors, a space interferometer is in development, which will make it possible to detect a new range of long-period gravitational waves in the later 2030s. Such longer waves should be abundant throughout the universe, caused by objects such as white dwarfs and massive black holes. Even longer gravitational waves, caused by the most cataclysmic events such as the collision of supermassive black holes, are expected to be detected with a different method known as pulsar timing.

Acknowledgments

I am very grateful to Maura Noone and Katherine Dooley for carefully editing the English version of this book. For many suggestions and patient proofreading, I thank very much the following individuals: Stefan Rottmann, Bettina Grote, Stefan Borchers, Walter Winkler, Peter Aufmuth, Katherine Dooley, and also Stefan Bollmann of publishing house C.H. Beck.

Many thanks to Josh Field for five illustrations, including the cover illustration. I also would like to thank my colleagues, particularly in the GEO and LIGO Scientific Collaborations, for decades (in some cases) of work together.

CHAPTER 1

They exist, they don't exist, they exist

1.1   GRAVITY: FROM NEWTON TO EINSTEIN

Gravity is considered the weakest of the physical forces, yet falling from a ladder onto the driveway would not be good for my health. When impacting the ground, it would be electromagnetic forces, the repelling of positive and negative charges, that would prevent my further falling through the asphalt: My body cannot easily penetrate it. The energy that would deform my body, however, is due to the gravitational force.

In addition to gravitational and electromagnetic forces, there are strong and weak nuclear forces that determine the stability and decay of atomic nuclei; however, they only act within a tiny distance from the nucleus. Although electromagnetic forces have a longer range than the strong and weak nuclear forces, they cancel out very quickly further away from the nucleus due to the equal amounts of positive and negative charges. At longer distances, gravitational forces dominate, creating an attraction between every kind of matter and energy known to us. In space, gravitation is the primary type of force that determines planetary motion, the life cycle of stars, and the evolution of the entire universe.

In the late seventeenth century, Isaac Newton formulated his laws of motion. With his meticulous quantitative analysis of how objects, or bodies of matter, move under the influence of force, he enabled unprecedented precision in the calculation of planetary orbits. In addition to these laws of motion, Newton also formulated a law of gravity that describes an attractive force between objects. According to this law, two bodies of matter attract each other with a force proportional to the mass of the two objects and inversely proportional to the square of the distance between them. Newton's gravitational force works immediately and without delay, that is, instantaneously. Under this theory, even when I type on a keyboard or lift a glass, this is swiftly communicated to the whole universe, because when objects change their position in space, the direction and strength of their outgoing gravitational force changes.

If we sum up Newton's law of gravitation and his laws of motion, we have the following chain of cause and effect: objects, such as our sun and the planets, exert instantaneously attractive forces which determine the way in which these objects move in space.

While this is the way Newton's laws describe how gravity works, Newton himself was doubtful to a degree about this aspect of his theory. In a letter to Richard Bentley from 25, February 1693 Newton writes: *It is inconceivable that inanimate Matter should, without the Mediation of something else, which is not material, operate upon, and affect other matter without mutual Contact. ... That Gravity should be innate, inherent and essential to Matter, so that one body may act upon another at a distance [through] a Vacuum, without the Mediation of any thing else, by and through which their Action and Force may be conveyed from one to another, is to me so great an Absurdity that I believe no Man who has in philosophical Matters a competent Faculty of thinking can ever fall into it. Gravity must be caused by an Agent acting constantly according to certain laws; but whether this Agent [is] material or immaterial, I have left to the Consideration of my readers.*

Regardless of this uncanny aspect of Newton's theory, during the eighteenth century, with his precise predictions of planetary constellations that were repeatedly confirmed by observations, Newton's theory became a triumph and it epitomized the power of the human mind.

In the early nineteenth century, however, a deviation of Uranus (the next planet beyond Saturn) from its calculated orbit had been observed. If Newton's theory was correct, there was only one good explanation for this deviation — there must be another, previously unknown, planet influencing the orbit of Uranus. Independently of one another, the French mathematician and astronomer Urbain

Jean Joseph Le Verrier and the Englishman John Couch Adams calculated the position of the unknown planet. Le Verrier asked Johann Gottfried Galle, an astronomer at the Berlin Observatory, to search in the section of the sky where his calculations had indicated its location. A short time later, Galle and his co-workers discovered the planet. Galle wrote to Le Verrier, *Monsieur, the planet whose position you have calculated actually exists!* A new planet had been found, for which Le Verrier later suggested the name Neptune.

The fact that a planet was discovered by a mathematical prediction was, once again, a grandiose confirmation of Newton's theory of gravitation. However, there was another problem — the innermost planet in the solar system, Mercury, also showed a deviation from its Newtonian calculated orbit. After each orbit around the sun, the perihelion, the point in the Mercury track when it is closest to the sun, will shift, moving a little further in space. In one hundred years this shift adds up to 574 arc seconds. (An arc second is one part in 3600 of a degree.) In Newtonian theory, most of this shift could be explained by the influence of the other planets. However, there remained almost 8 percent (45 arc seconds) that were unexplained. After his triumphant prediction of the existence of Neptune, Le Verrier was now convinced that this anomaly of the Mercury track was caused by yet another unknown planet called Vulcan. It was a big mystery as to why this planet that must have its orbit so close to the sun had not already been observed.

More than fifty years later, only Albert Einstein was able to solve this riddle. In 1905, Einstein proposed a new theory of space and time, which follows from two assumptions: (1) Light always travels at the same speed, regardless of the speed of the source of the light or the speed of the observer. (2) The laws of physics in uniformly moving reference systems (systems that are not accelerated, also called inertial systems) are always the same; this is the principle of relativity formulated by Galileo. From these assumptions, the Special Theory of Relativity emerged, resulting in a close intermeshing of space and time, expressed by the concept of space-time. One consequence of the Special Theory of Relativity was the concept that nothing, including information, could travel faster than the speed of light. Among other considerations, this radically new idea led Einstein to the conclusion that not only space and time, but also Newton's law of gravitation required a revision because according to Newton, the gravitational force spread instantaneously and with infinite speed. Up until now, instantaneous effect of gravity at any distance had not been called into question by most physicists. However, in light of the Special Theory of Relativity, it was no longer conceivable, and Einstein set to work to develop a new, more compatible theory of gravitation that would eventually become the General Theory of Relativity. Special Relativity is called 'special' because it only deals with inertial systems and in particular does not include a description of gravity. The new theory of gravitation, on the other hand, was given the name General Relativity because it also describes gravitation.

For Einstein, the central, motivating factor in the development of a new theory of gravitation was the apparent indistinguishability of acceleration and gravitational attraction. An astronaut in a windowless rocket has no way of knowing if she is sitting in the stationary rocket on the ground, waiting for takeoff, or if she is in an accelerating rocket in interstellar space. In both cases, her body would be pressed into the seat with the same force. Admittedly, this is a somewhat bizarre example, because an astronaut would probably always know where she is, but it is typical of the thought experiments that Einstein often employed. The indistinguishability of acceleration and gravitational attraction is also called the (strong) principle of equivalence, and Einstein called this idea *the happiest thought of my life.*

After several years of laborious work, Einstein's thought experiments eventually led to the General Theory of Relativity, published on November 25, 1915. Figure 1.1 shows the beginning of this article. The hallmark of the new theory is that space itself must be considered deformable, whereas previously it had been considered immutable and flat, at least by Newton. More precisely, not only is space deformable, but space-time is deformable. (The unity of space and time previously introduced by the Special Theory of Relativity.) The potential deformation of time sounds very strange. This refers to a dilation of time, which means, for example, that clocks in a curved space do slow down. As a result, we simply often speak of the curvature of space and ignore the role of time, with the awareness that time is always dilated in curved space. What causes the curvature of space? Mass in the form of either matter or energy. Both the sun and an apple bend the surrounding space. Since the mass of the sun is larger than that of an apple, the curvature caused by the sun is also larger.

The curvature of space, caused by matter, results in all objects in this space experiencing a force in the direction of the curvature. In this theory, space serves as a kind of mediator and carrier of

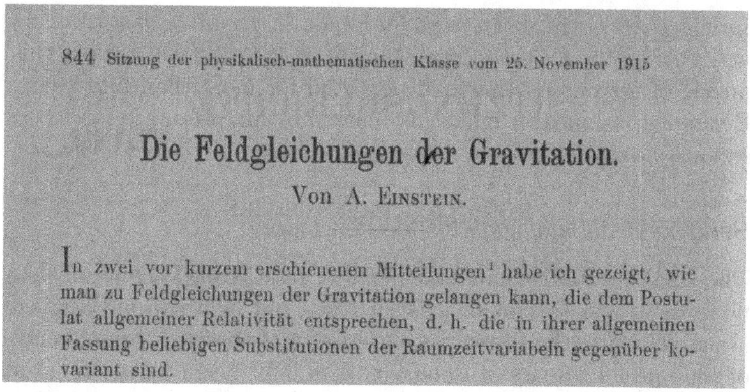

844 Sitzung der physikalisch-mathematischen Klasse vom 25. November 1915

Die Feldgleichungen der Gravitation.

Von A. EINSTEIN.

In zwei vor kurzem erschienenen Mitteilungen[1] habe ich gezeigt, wie man zu Feldgleichungen der Gravitation gelangen kann, die dem Postulat allgemeiner Relativität entsprechen, d. h. die in ihrer allgemeinen Fassung beliebigen Substitutionen der Raumzeitvariabeln gegenüber kovariant sind.

Figure 1.1   The beginning of Einstein's famous article about the field equations of General Relativity. The paper was published in the proceedings of the Royal Prussian Academy of Sciences on November 25, 1915.

information between objects and thus instantaneous action at a distance is no more required. This is in contrast to Newton's theory where space does not play such a role.

To summarize, in the General Theory of Relativity we have the following chain of cause and effect: matter or energy curve space and the curved space determines the way objects move. In the catchy words of the physicist John Archibald Wheeler: *Matter tells space-time how to curve and space-time tells matter how to move.* Gravitation is contained in the geometry of space itself. This structure is also reflected in the Einstein field equations, which state that mass (or energy) and space curvature are directly related.

In the early stages of the development of his new theory, Einstein, together with his friend Michele Besso, calculated the effect of the sun-curved space on the perihelion shift of Mercury's orbit. In another November 25, 1915, publication, he writes: *The calculation provides for the planet Mercury a progression of 43" (arc seconds) in a hundred years, while astronomers specified 45 (±5)" as an unexplained remnant between observations and Newtonian theory. This means full agreement.* Einstein's new theory yields the value that was actually observed by astronomers, for which there had previously been no explanation! A short time later, Einstein notes in a letter to Arno Sommerfeld on December 9, 1915: *The result of the perihelion shift of Mercury fills me with great satisfaction. Great to see how the pedantic accuracy of astronomy helps us, which I used to make fun of in the past!* By solving this puzzle of astronomy that had persisted for more than fifty years, Einstein continues to make his mark and his peers are forced to seriously consider his new theory.

Bending of starlight confirms the new theory

The smallest energy unit of a beam of light, called a photon, has no mass at rest (it cannot exist at rest), but light can be assigned a mass through the equivalence of mass and energy, a consequence of the Special Theory of Relativity. If a light beam passes another object, like a star, the beam is slightly deflected. The Special Theory of Relativity allowed for this prediction, due to the attraction between objects demanded by Newton's law of gravitation. In the November 25, 1915, publication, Einstein corrected this earlier prediction of light deflection; he found that the deflection of light is twice as large using the General Theory of Relativity.

Some time after this prediction, it became clear that the bending of starlight could possibly be observed in the vicinity of our sun. Comparing the apparent positions of stars that are visible close to the sun with their apparent positions when viewed at positions in the sky far away from the sun would reveal the bending of light if those positions were found to be different. A total solar eclipse is required for this method, in order to be able to see any stars visible in the vicinity of the sun. Figure 1.2 shows a cartoon of the bending of starlight at the sun.

The total solar eclipse of May 29, 1919, seemed particularly suitable for this measurement, since the sun would be in front of a group of bright stars, called the Hyades. In two famous expeditions to observe this eclipse, the Briton Sir Arthur Stanley Eddington and his colleagues measured the deflection of starlight by the sun. To increase the chance of success, one team, led by Eddington, ventured to the small island of Principe, off the coast of west Africa, whereas the other expedition, led by Andrew Crommelin, was sent to Sobral in northern Brazil. Figure 1.3 shows the telescopes of this expedition, set up in Sobral.

On the day of the eclipse, the weather was mostly cloudy in Principe, and only just when the total eclipse occured, allowed the clearing of clouds Eddington's team to take two usable photographs. Furthermore, the group had to leave earlier than planned and had no time to take reference photographs of the star

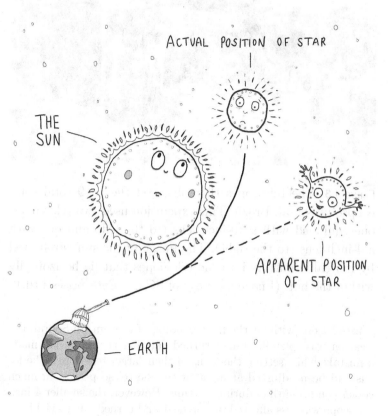

Figure 1.2 The bending of light at gravitating objects like the sun leads to a shift in the perceived position of a star. (Illustration: Josh Field.)

Figure 1.3 The telescopes that observed the 1919 total solar eclipse in Sobral, Brazil. This expedition used two telescopes, one equipped with a 10-inch lens (to the left) and one with a 4-inch lens (to the right). The image of the sun is reflected by moveable mirrors into the telescopes that lie horizontally within the hut. (Image courtesy of SCMG Enterprises Ltd.)

constellations without the light bending of the sun. In Sobral, the weather was splendid, but it turned out that the team had made a mistake while setting the focus of their larger telescope. The focus had been adjusted at night, when the telescope was in much cooler condition than during daytime. However, the smaller 4-inch telescope was less affected by the heat and turned out to yield the best photographs of both expeditions.

Back home, the data is being analysed thoroughly and the result is announced on a dedicated meeting of the Royal Astronomical Society and the Royal Society of London on November 6, 1919. In spite of the fact that the measurement inaccuracy is not insignificant, the measured value is closer to that predicted by the General Theory of Relativity. For light rays that pass directly on the sun's edge, the Sobral expedition obtains a value of 1.98 ± 0.16 seconds of arc. The value obtained by the Principe expedition is less convincing and obtains a value of 1.61 ± 0.40 seconds of arc.

However, both values are closest to the value predicted by Einstein of 1.74 seconds of arc, which is about one-thousandth of the angle at which the sun's diameter appears as seen from earth.

The day after the announcement, the *London Times* features an article titled: *REVOLUTION IN SCIENCE. NEW THEORY OF THE UNIVERSE*. And two days later, *The New York Times* appears with the headline: *LIGHT ALL ASKEW IN THE HEAVENS*.

The confirmation of his prediction of light deflection instantly makes Einstein a world-famous physicist, not only within the professional world, but also in the public. He becomes somewhat of a science pop-culture icon.

Other predictions by General Relativity

Another early prediction emerging from the General Theory of Relativity concerns the influence of matter on the passage of time. As previously mentioned, time in a curved space near an object, like the surface of a star, passes more slowly. Here, again, we have the close integration of space and time. Since light has a certain frequency of oscillation, one can compare the emission of light with a clock that is ticking at a particular frequency. One could say that in the curved space on the surface of the star, an atom emitting light is ticking a little slower than it would in an uncurved, flat space. This lower oscillation frequency causes the emitted light to have a slightly longer wavelength. We call this light red-shifted, because red light has a longer wavelength than other colors of light visible to the human eye. Although the effect of this gravitational redshift is small—the wavelength change for light emitted by our sun is about 0.0002 %—it was already indirectly measured in the 1920s.

Thus, three consequences of general relativity could be confirmed within a few years of its publication in 1915. These are the anomaly of the perihelion shift of Mercury, the amount of light deflection by the sun, and the gravitational redshift. The confirmation of two additional consequences of general relativity took considerably longer. These are the existence of black holes and the existence of gravitational waves. Although the two are not necessarily related, two particular black holes merged 1.4 billion years ago emitting the first gravitational waves to be detected on earth. Chapter 6 will convey this story.

As early as 1915, the German astronomer and physicist Karl Schwarzschild found a solution to Einstein's equations describing a singularity of space-time. This was the first indication of something that would eventually be coined by the term "black hole". A black hole is an area of very strongly curved space characterized by an extremely dense accumulation of mass. The mass is concentrated in such a small area that the gravitational redshift reaches infinity, as one gets closer and closer to the black hole. Near the so-called event horizon of the black hole, clocks would appear (if observed from far away) to tick almost infinitely slowly. The gravitational redshift approaches infinity as well, and thus no light can escape from the event horizon or beyond. This property eventually inspired the term "black hole", proposed by the physicist John Archibald Wheeler, although it had previously appeared occasionally in the scientific community.

Over the decades since 1915, the evidence that black holes actually occur in nature and contribute significantly to the dynamics and development of the universe has become stronger and stronger. Particularly impressive is the observation that in the center of our galaxy, the Milky Way, some stars circle very tightly around an invisible center. Currently, the only scientifically plausible explanation for this phenomenon is the existence of a black hole in this invisible center, with a mass equivalent to that of about four million suns. And in April 2019, the Event Horizon Telescope collaboration revealed the first image of a supermassive black hole located in the center of the Galaxy M87 (see Chapter 7).

1.2 THE PREDICTION OF GRAVITATIONAL WAVES FROM GENERAL RELATIVITY

In Newtonian theory, gravitation has an instantaneous effect. Since a wave consists of the traveling of a disturbance in space and time, there can be no waves of gravitation, according to this theory. In the late eighteenth century, the mathematician and physicist Pierre Simon de Laplace had already begun to consider what would happen if the gravitational force did not travel instantaneously but at a finite speed. At this time, these considerations were related to the acute problem of describing the orbit of the earth's moon as precisely as possible with Newtonian theory. Although Newton's theory had been so successful in calculating the planetary orbits, it had been very difficult to apply it equally successfully to the orbit of the earth's moon, as the moon is subject to many additional influences. However, Laplace's investigations determined that Newton's theory was sufficient. The speed of propagation of the gravitational force had to be at least 100 million times the speed of light. This value is certainly close to infinite speed, at least in a physical context. For many years, this result halted speculation about a finite speed of propagation and the instantaneous effect of the gravitational force was widely accepted, for the time being.

Around 1905, the idea of gravitational waves emerged once again, with Henri Poincaré. The Special Theory of Relativity proposed that no information can travel faster than the speed of light. According to Poincaré, if the gravitational force would travel at finite speed, waves of gravitation would be possible. Poincaré applied this idea to the problem of the anomaly in the perihelion shift of Mercury, but came to the conclusion that the emission of gravitational waves by Mercury could not be a sufficient explanation for its orbital anomaly. As previously mentioned, Einstein was going to be the first to explain this anomaly with the curvature of space.

With the General Theory of Relativity of 1915, the idea of gravitational waves became more concrete. Given the complexity of the equations of general relativity, Einstein recognizes gravitational waves as a possible solution to them only some time later, in 1916. Generally speaking, it is a task of theoretical physicists to find possible solutions to these ten, nonlinear equations. This is analogous to solving puzzles and there are many possible solutions; however, not all solutions necessarily have a meaningful physical reality. To master the complexity of the equations, Einstein and his colleagues used approximation techniques. In addition, they had to choose a particular coordinate system in order to calculate solutions for the problem described. In principle, all coordinate systems should yield the same results, but different coordinate systems are better suited to address certain questions. The choice of the coordinate system in combination with the approximation methods initially led to errors in the interpretation of the results. This is the reason that Einstein did not initially believe in the existence of gravitational waves in 1915, but he changed his mind in the middle of 1916. The astronomer Willem de Sitter had pointed out to him that another coordinate system was more suitable for studying the existence of gravitational waves. Further, Einstein introduced a linearization of the otherwise nonlinear equations, resulting in an analogy with the equations of electromagnetic waves. A linearization of an equation means to simplify the equation and make it valid only in an approximate sense and for sufficiently small amplitudes of the quantities described by the equation.

After experimenting with different coordinate systems and predicting either three different forms of gravitational waves or just one, Einstein concluded that there must be only this one form of gravitational waves. Due to the choice of coordinates, the other two would be artifacts, meaningless by-products of the calculations. However, it seems fair to assume that Einstein believed that even if gravitational waves existed, it would never be possible to measure them. In 1916, at the end of his calculation of the strength lambda (Λ) of gravitational waves, Einstein writes: *So one sees that Λ in all conceivable cases must have a practically vanishing value.* Apparently, Einstein thought the waves were just too small to be ever detected. At the time, this was a reasonable conclusion because the existence of compact, heavy celestial objects, such as black holes or neutron stars, was not yet known, nor were many of the technologies that would allow the first measurement of gravitational waves one hundred years later.

Just the recognition of the existence of gravitational waves was a difficult birth and, as it turned out later, Einstein had made a mistake in his work of 1916 due to an incorrect approximation regarding how much energy is contained in a gravitational wave. It took him two years to correct this mistake. It was a Finnish physicist, Gunnar Nordström, who recognized the inconsistency in Einstein's gravitational wave result of 1916 and communicated it to him in the fall of 1917. In early 1918, Einstein published an essay in which he corrected his mistake and introduced a new formula for the emission of gravitational waves.

This new formula demonstrates that gravitational waves are emitted by accelerating matter, which is matter that changes its speed or the direction of motion in space. However, there is a limitation inherent in the formula — not all accelerating matter emits gravitational waves. For example, a completely spherically symmetric explosion of a star will not send out gravitational waves. In this case, all partial waves would cancel each other out. Einstein's formula for radiating waves requires an asymmetry, which, for example, occurs when two masses rotate around each other.

The next physicist to focus on gravitational waves was Sir Arthur Stanley Eddington. In his view, it was an unresolved question at which speed gravitational waves spread. Analogous to electromagnetic theory, Einstein had assumed that gravitational waves would travel at the speed of light. Eddington, however, was initially not convinced of the applicability of this analogy and acted as a profound skeptic of gravitational waves, a notion still in its infancy. His famous statement that gravitational waves traveled *at the speed of thought* is often cited as an indication of his doubts about their existence. In fact, Eddington's statement only referred to those waves from Einstein's equations determined to be mathematical artifacts in the application of certain coordinate systems. In a 1922 essay, Eddington re-derives the formula for the emission of gravitational waves and corrects another small error in Einstein's 1918 essay.

Through his investigations, Eddington was able to convince himself that gravitational waves actually did travel at the speed of light. Next, he turned his attention to the question of what impact the emission of the waves would have on the emitting source, or the object emitting the waves. The formula for emitting gravitational waves states that two masses orbiting each other, such as two stars, will produce gravitational waves and, as a result, will lose energy due to their emission. This loss of energy causes the stars to approach each other and begin to orbit faster. The cycle ends only when they have come so close that they practically collapse and merge into a single star, or object. At the time, many physicists and astronomers had difficulty with this scenario because of the so-called two-body problem. Previously, the orbiting of two celestial objects as a binary system was considered stable and was a solid foundation of celestial mechanics. In other words, two objects orbiting each other should do just that, for all time. The idea that they might collapse meant a fundamental instability in the constellation. The controversy surrounding the two-body problem persisted for some time until researchers became convinced that binary systems, like stars orbiting around each other, actually do lose energy through the emission of gravitational waves and are, thus, unstable. However, the energy loss is so small that it only plays a significant role in the case of very compact and massive objects, such as black holes, that circle at a small distance with large velocity.

In 1936, there was a curious intermezzo in the form of a work that Einstein had submitted for publication in the prestigious journal *Physical Review*, entitled, "Do Gravitational Waves Exist?". In a letter to Max Born, Einstein writes: *Together with a young colleague (Rosen), I found the interesting result that there are no gravitational waves, even though this was, according to the first approximation, considered to be the case for sure.* In an interesting twist, the editor of the *Physical Review* would not publish Einstein's work because an anonymous reviewer believed his conclusion to be wrong. From his experience with German journals, Einstein is unaccustomed to the anonymous review process and writes a furious letter to the *Physical Review*, withdrawing his work. A few months later, Einstein recognizes his mistake and publishes the article with a different title in another magazine. He now considers it likely that gravitational waves really exist. Generated by accelerated objects that stretch and compress the surrounding space, they are ripples of space-time, traveling at the speed of light.

1.3 ASTRONOMICAL CAUSES OF GRAVITATIONAL WAVES

In principle, even moving objects in our everyday world, such as the running engine of a car, generate gravitational waves. However, those waves are much weaker than the waves generated by massive objects throughout the universe. For this reason, gravitational wave research is intimately linked with astronomy.

Einstein, after deriving the formula that describes the amplitude (or the strength) of gravitational waves emitted by accelerated objects, recognized that the amplitude depends on the masses of the accelerated objects and even more so on the magnitude of their acceleration (their change in speed, let it be magnitude or direction). Even two ordinary suns that circle very closely emit only very weak gravitational waves since their acceleration is small. It was soon clear that astronomical objects as heavy as ordinary suns, but much smaller and more compact, were required in order for there to be any chance of observing gravitational waves with instruments on earth. Such compact objects can orbit much faster on narrow paths, thus producing much stronger waves.

To satisfy the requirement for compact astronomical objects, if one has the goal to detect gravitational waves, neutron stars and black holes come into play. A neutron star is a compact remnant of a star that has completed its active phase as a sun and collapsed in a powerful explosion called a supernova. In order for a supernova to form a neutron star, a sun must have a particular mass before collapsing (not too heavy, but not too light). The name for a neutron star has been chosen because it essentially consists of atomic-nucleus matter in the form of neutrons. A neutron star has about the same mass as our sun but is one hundred thousand times smaller in diameter. Neutron stars are extreme forms of known matter. A neutron star that gets too heavy, for example by accreting more mass from its surrounding, will ultimately collapse into a black hole. This also means that if the mass of the original sun, or star, exceeds a certain limit, then a black hole is formed in a supernova instead of a neutron star. Since about half of all stars exist in binary systems, that is, have partner stars around which they circle, many compact binary systems emerge at the end of the lifetime of the stars, consisting of two black holes or two neutron stars. Another way to form compact binary systems is by a close encounter, so to speak, of three compact objects in space. If this happens, the laws of dynamics govern resulting in two of the compact objects to form a binary system. While two objects thus continue to orbit around each other, the third one is ejected from the encounter with sufficient velocity to escape the scene.

Currently, physicists and astronomers anticipate four distinct types of sources of gravitational waves that can be identified within the frequency range of terrestrial gravitational-wave detectors. They are: compact binary objects, supernovae, rotating neutron stars and background noise. It is known that compact binary systems can consist of two neutron stars, a neutron star and a black hole, or two black holes. As previously explained, the binary systems lose energy through the emission of gravitational waves, decreasing the distance between the objects and increasing their orbiting speed. The increase of speed as a result of losing energy may seem strange and is called the orbit paradox. It is explained though by taking into account the potential energy loss from the closer orbit. As objects circle one another with increasing speed, gravitational waves of increasing frequency and strength are generated until the two objects finally merge into one.

In the brief moment of the explosion of a star, supernovae generate gravitational waves provided that there is an asymmetry in the acceleration of matter in the explosion. The magnitude of this asymmetry and thus the amplitude and waveform of the resulting gravitational wave is not well known and is the subject of active research, mainly carried out with computer simulations.

Similar to supernovae, neutron stars that rotate around their axes are also a source of gravitational waves, provided they have an asymmetry in their mass distribution. In a nutshell, this is the case if they have a hump. In contrast to the short impulses expected from a supernova, the generation of gravitational waves by such rotating neutron stars takes place continuously. The observation of radio pulsars (see Chapter 7) shows that some neutron stars orbit their own axis up to several hundred times per second.

Finally, gravitational waves can also be expected in the form of noise that cannot be clearly assigned to individual sources. This noise can have various causes, such as the unresolved superposition of many individual sources of gravitational waves, or even of gravitational waves generated immediately after the Big Bang.

Figure 1.4 summarizes the four expected sources of gravitational waves.

From top to bottom: Two compact massive objects, like black holes or neutron stars, inspiralling into each other; a star exploding in a supernova; spinning neutron stars with a small mass asymmetry; a multitude of individual inspiralling sources, combining to a stochastic background of gravitational waves.

The astronomical objects and events listed here are the most important causes of gravitational waves currently known in theory. Of course, there is hope to identify more. Gravitational wave impulses of unknown form could lead to the discovery of completely new astrophysical processes. But let us now turn to the question of how gravitational waves can be measured.

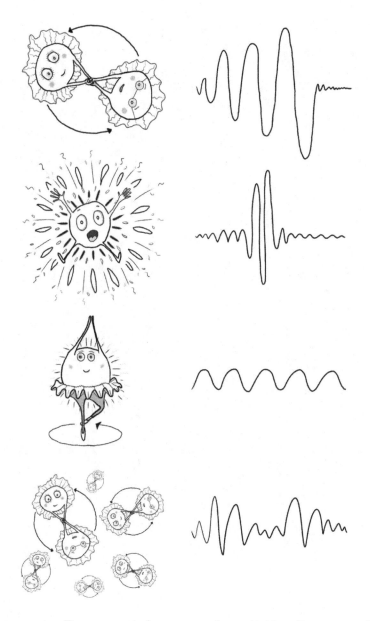

Figure 1.4 Four expected sources of gravitational waves and their corresponding gravitational wave-forms. (Illustration: Josh Field.)

They exist, they don't exist

In the late 1950s and early 1960s, the existence of gravitational waves was largely agreed upon, although it was still unclear whether or not they contained energy and interacted with matter. The question of whether a detector could ever be constructed to measure them was still being debated. In a Polish journal in 1956, the physicist Felix Pirani had published an essay which addressed this question but the article received little attention at the time.

Initially, not much was known about the type of events and astronomical objects that could generate significant gravitational waves. Although the existence of black holes and neutron stars was recognized as theoretically possible, it was not until 1967 that astronomers were able to confirm the existence of a neutron star (see Chapter 7). For this reason, supernovae appeared to be the best candidates for emitting gravitational waves, but there were no reliable estimates of the strength of the waves they would generate.

As Einstein had already recognized, the deformation of space by gravitational waves is extremely small compared to the dimensions of our everyday world. Space-time is deformable, but it is extremely rigid, which means that in order to deform it, an enormous amount of energy is required. Gravitational waves are strongest when closest to their source and are, therefore, easiest to measure for nearby events. The amount of deformation of space decreases proportionally to the distance from their source. This means that the immense size of the universe and the spatial and temporal distribution of events that generate gravitational waves are central factors in estimating the sensitivity of a detector that is required to make a successful measurement.

2.1 JOSEPH WEBER

The history of the experimental search for gravitational waves begins with the American Joseph Weber, born in 1919, the son of Jewish immigrants. During the Second World War, he not only escaped the sinking of the aircraft carrier Lexington, but commanded a submarine fighter and temporarily worked in reconnaissance. After the war, in 1948, he became a professor of electrical engineering at the University of Maryland and a professor of physics in 1950. He is considered one of the few scientists who excelled in both the experimental and theoretical domains of physics. In the early 1950s, Weber considered the possibility of building a laser, which wouldn't be achieved until a decade later. A laser (acronym for Light Amplification by Stimulated Emission of Radiation) is a special light source that can produce extremely monochromatic light, that is, light of a single color. In 1917, the possibility of stimulated emission of light had already been recognized by Einstein. Lasers are necessary for the operation of modern interferometers, and we will discuss them again in the next chapter.

In 1952, Weber gave a lecture at a conference in Ottawa, Canada, on the possibility of amplifying electromagnetic radiation, which led to a publication the following year. Weber was not the only physicist working on this topic though. In 1954, Charles Townes from Columbia University and his co-workers built the first maser (a variant of the laser for longer-wave radiation). The Russian physicists Nikolay Basov and Aleksandr Prokhorov made contributions to the theory in 1954 and 1956. After Theodore Maiman had constructed the first laser at Hughes Research Laboratories in 1960, Townes, Prokhorov and Basov received the Nobel Prize in Physics in 1964 and Joseph Weber, despite his contributions, was left out in the cold.

During this time, Weber's attention was focused on a new research interest — gravitation and gravitational waves. In 1957, he attended a famous conference in Chapel Hill, North Carolina, financed by private sponsors who wanted to advance research into gravity. Inspired by Felix Pirani's work, consensus was formed at this conference, that gravitational waves could, in principle, indeed be measured. Weber, who also was inspired by John Wheeler, wrote an essay in 1960 on the theory and detectability of gravitational waves. He concluded that it would be technically possible to measure gravitational waves and he single-mindedly got to work. Wheeler later wrote in his autobiography: *He [Weber] threw himself with religious fervor at the gravitational waves and pursued them for the rest of his career. Sometimes I wonder if I didn't fill him with too much enthusiasm for this monumental task.*

Since gravitational waves stretch and compress space, Weber's basic idea was to use a massive object that would be stimulated to oscillations by a passing gravitational wave. He chose to use a solid cylinder of a size that could just be handled conveniently in a laboratory. The cylinder would be made of an aluminium alloy, and be suspended at its center with wires, allowing it to swing and oscillate freely. The vibration of the so-called longitudinal mode is important to note. This is the mode of vibration of a cylinder in which its flat end surfaces move parallel, but in opposite directions to each other. When a gravitational wave passes through the cylinder, the cylinder experiences an expanding and compressing force, which is caused by the expansion and compression of space-time by the wave. If the excitation is large enough, the cylinder oscillates for a short time with its own natural frequency; this is roughly comparable to a bell, which oscillates for a certain time with its own particular sound. With the proper techniques, the vibration of the cylinder is then converted into electrical signals and subsequently recorded.

Technically, a gravitational wave (like any form of excitation) not only causes an increase but can also cause a decrease in the amplitude of the vibration of the cylinder! This is because the amplitude normally fluctuates around an average value, which means that it can also become smaller if the excitation is in the appropriate phase. For this reason, when searching for gravitational waves with cylinders, it is best to search for *changes* in the vibration amplitude.

The excitation of a cylinder by a gravitational wave works most effectively when the oscillation frequency of the space-time stretching corresponds to the natural frequency of the cylinder, meaning that the cylinder is sensitive to a particular frequency of gravitational waves. These cylinders, sometimes also called Weber cylinders, are referred to as resonant antennas, since they are sensitive to gravitational waves in the range around a certain frequency, i.e., resonant. The term *antenna* is associated with the reception of electromagnetic waves, but the function of a Weber cylinder is to receive gravitational waves. The term detector is often used as a general synonym for a measuring instrument, such as the Weber

Figure 2.1 Joseph Weber at one of his cylinders in Maryland. Visible in the middle of the cylinder are square plates on the surface, called piezoelectric elements, that convert the strains and compressions of the cylinder into electrical signals. (Image: Special Collections and University Archives, University of Maryland Libraries.)

cylinder. Figure 2.1 shows Joseph Weber at work. Weber's early cylinders were sensitive at a frequency of 1660 Hz. Largely to shield the cylinder from ground vibrations, the device was mounted on alternating layers of rubber and metal (bottom left in Figure 2.1). In addition, the cylinder is operated in a vacuum chamber (not visible in Figure 2.1) to protect it from acoustic influences and temperature fluctuations.

The sensitivity of a gravitational wave detector is generally given as a measure of the relative strain, or expansion, of space which can just barely be registered as distinct from the noise of the detector. The relative strain of space corresponds to the absolute change in length divided by the length of the measuring section. Weber's cylinders, which were operated at room temperature, achieved a sensitivity of about 10^{-16} m at an oscillation frequency of 1660 Hz. This means that his detectors were able to measure a length change of one-tenth the diameter of a proton over a length of one meter. A relative change in length of 10^{-16} roughly corresponds to a change in distance between the earth and the moon of just the diameter of an atom, or to the change of a hair's breadth between the sun and the earth. And this was just the beginning!

As a technical note, the detector's sensitivity specification always includes the duration of a measured signal. In other words, besides the main frequency, the sensitivity always refers to a bandwidth of the detector, the range of frequencies that can be measured around the central frequency. Without further information, a bandwidth of one Hertz is assumed, which corresponds to a signal duration of one second.

2.2 HOW TO DISTINGUISH SIGNALS FROM NOISE?

A fundamental problem when measuring weak signals such as gravitational waves is how to reliably distinguish them from the natural noise of the measuring instrument. Inherent to each physical measuring apparatus is some internal noise which limits the sensitivity of the device. In physics, noise is often a process that is ultimately based on random, but statistically calculable, fluctuations that are homogeneous over time. Thermal noise caused by the statistical movement of the atoms and molecules that make up the cylinder is one factor that limits the sensitivity of a Weber cylinder. The thermal kinetic energy, which is equivalent to the temperature of the cylinder, will consistently excite the cylinder to oscillations at its resonance frequency.

A change in the vibration state of the Weber cylinder by a gravitational wave can only be distinguished from the noise of the detector if the change is sufficiently large. It is important to note that in the analysis and interpretation of the measurements there are no clearly defined limits to distinguish a signal from noise.

Although the evaluation is largely automated at the present time, each individual case of a fluctuation in the output signal of the detector must be evaluated to determine whether or not it could be a desired signal. Ultimately, this evolves into a statistical analysis and a probability statement about the possible cause of an observed signal.

In addition to the sources of internal noise inherent to the detector, events from the external environment can generate signals (usually for a short time) in the measuring apparatus that can also be misinterpreted to be the sought-after signals. These disruptive signals can be caused by circumstances such as ground vibrations caused by small earthquakes, road traffic or wind, or by electromagnetic disturbances. Much of the experimental work involved in setting up the detectors consists of shielding them as best as possible from disturbing environmental influences. Shielding methods include seismic isolation systems and vacuum chambers. Since these methods are never quite perfect and there can also be the internal disturbances inherent to the detector, two basic techniques are utilized to distinguish a sought-after signal from a disturbance.

The first technique is to employ the most appropriate and highly sensitive sensors in the surrounding environment. For example, a seismometer would be used for recording ground vibrations and a different specialized sensor would be used for detecting magnetic field fluctuations. If a signal in one of these environmental sensors occurs simultaneously with a signal in the measuring apparatus, it is assumed with a conservative, cautious attitude that the alleged signal in the measuring apparatus was caused by a disturbance from the environment. This principle is called vetoing — the signal in the environmental sensor serves as a veto when interpreting the signal in the detector. The caution required in interpreting the measurement data is due to the novelty of this field of research. Should a signal not be linked to a disturbance in the environment, then the alternative hypothesis would be the measurement of a gravitational wave! This would have been considered a particularly strong claim, given the fact that it would have been the first such measurement.

The second technique is to employ the principle of coincidence measurement in which a detection of the desired signal is only assumed if it has been registered in at least two detectors simultaneously within a short period of time. If two detectors are sufficiently far apart, it is unlikely (though not completely excluded) that an interference will occur at the same time. A fundamental question in coincidence measurements is how many random, or unwanted, coincidences are obtained naturally, i.e., without the presence of a sought-after signal such as a gravitational wave. From this knowledge, one can then deduce with what probability an observed coincidence actually points to a gravitational wave and not a signal caused by external or internal disturbances. We will discuss this topic in more detail in Chapter 5.

To illustrate the principles of veto and coincidence, Figure 2.2 shows simulated signals from two detectors and their environmental sensors.

2.3 CONTROVERSY AND CONSENSUS

Between 1967 and 1970, Weber published five articles in the renowned journal *Physical Review Letters* and, with increasing certainty, he announced the probable detection of gravitational waves. While Weber initially stated that it cannot be ruled out that the signals he observed may originate from gravitational waves, his conclusion was more definitive in the 1969 publication. He writes: *This is good evidence that gravitational radiation has been discovered.* This latter paper, in which Weber compares data from two cylinders 1000 kilometers apart in a coincidence measurement, arouses the interest of researchers worldwide and they begin to request the details of Weber's measurements.

At the end of 1969, Robert Forward, a member of Weber's staff, contacted at least 13 colleagues from all over the world and offered technical support and advice to those interested in repeating Weber's experiments. Many independent laboratories began to embark upon building resonance antennas to verify Weber's claim.

Theoretically, it seemed implausible that there would be gravitational waves as strong as Weber believed he had observed. If these signals were from supernovae in our own galaxy, the Milky Way, it would mean that most of the stars in the Milky Way would already have had to have exploded. This is clearly not the case! Ultimately, it was necessary that the experiments of other research groups demonstrate whether or not Weber's results were reproducible, because, in principle, there could also have been errors in the calculation of the signal strengths or in the estimate of the actual sensitivity of the detectors.

In 1970, Weber presented two further publications in which he came to the same conclusion — that he had indeed measured gravitational waves. In the first publication, along with the analysis of

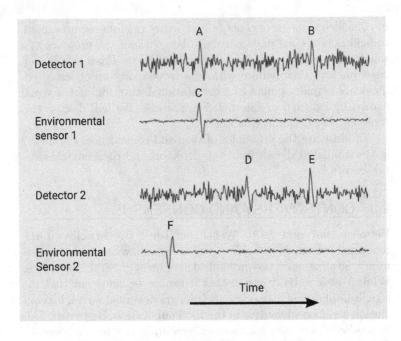

Figure 2.2 Simulated signals of two detectors and their environmental sensors. In detector 1, events A and B are marked — a visible structure stands out from the internal noise of the detector. Environmental sensor 1 registers an event C, which occurs almost simultaneously with event A. In the analysis, C is used as a veto against the interpretation of A as a sought-after signal. In detector 2, events D and E are obvious. Since no event occurs at the same time as event D in detector 1 or in environmental sensor 2, D can be classified as an internal disturbance in detector 2. Event E (detector 2) occurs almost simultaneously with event B (detector 1), making these events potential candidates for a sought-after signal, like a gravitational wave. The fact that not all environmental disturbances couple into the detectors is illustrated by Event F.

his data, he introduced a new concept — experimentally determining the rate of random coincidences. In order to achieve this, data is collected and analyzed from two detectors that were recorded at *different* times. The idea is that coincidences (simultaneous events in both detectors) found in such a time-shifted analysis are unlikely to be caused by a gravitational wave, since a gravitational wave would have registered almost simultaneously in both detectors. The time-shifted analysis gives an estimate of the rate of coincidences that *do not* contain a gravitational wave signal. This rate can then be compared to the rate of coincidences obtained when analyzing the detector data from the *same* time segments.

In the second publication, Weber writes that there are strong indications that the signals he registered came from the direction of the center of the Milky Way (the galactic center). Such an observation would demonstrate that the coincidences could indeed be caused by gravitational waves.

In 1972, the first experimental results by other research groups were published. Some are not as clear as others, but, overall, none of the groups using resonant antennas were able to reproduce any signals of the kind observed by Weber.

At the time, Weber's work was also being followed at the Max Planck Institute for Astrophysics in Munich, which became the nucleus of gravitational wave astronomy in Germany. Heinz Billing and his colleague, Walter Winkler, started to build a resonance antenna modeled as closely as possible after Weber's detector. The group also introduced substantial improvements that resulted in the most sensitive detector of its time. They optimized the piezoelements for reading the vibration amplitude, developed quieter electronics, and, when analyzing the data, employed the principle of searching for *changes* in amplitude and not only for increases in amplitude, as Weber had done. The Munich group analyzed the data of their resonant antenna at the same time as a group in Frascati, Italy, which had assembled a detector of very similar construction. The results were published in 1975 and 1977 — once again, there was no evidence of gravitational wave measurements.

The lack of successful replications increased doubts about Weber's findings and, in some instances, generated fierce controversy. Experimental results are always open to interpretation, but most scientists were becoming increasingly skeptical. Weber stood by his interpretation of the data and published again in 1972 and 1973. Although he had made some improvements to his data analysis, he also made mistakes. He erroneously discovered a significant number of coincidences between detectors whose data were recorded at different times and, therefore, were not likely to show any coincidences caused by gravitational waves at all. Every scientist can make mistakes, but in the context of the controversy surrounding

Weber's results, this heightened doubts about the credibility of his data analysis. Moreover, in Weber's new data, the galactic center is no longer apparent as the preferred direction, and in spite of the fact that his detectors became more sensitive over time, the strength of the observed signals did not increase accordingly.

The end result was that Weber was largely ignored by his fellow scientists from 1975 onward. Until his death in 2000, he continued to work on the detection of gravitational waves and was convinced of the reality of his measurements. Indisputable proof to the contrary is not possible, but the scientific consensus is that Weber was wrong. Nevertheless, he deserves recognition as having initiated the experimental search for gravitational waves. John Wheeler praised him, saying: *Weber's merit remains to have shown the way. Nobody had the courage to look for gravitational waves until Weber showed that it was possible.* Weber's dedication inspired many researchers to continue the search for gravitational waves. Some were turning toward the new technology of interferometers, while others were working on significant improvements to the resonant antennas.

2.4   AN EXCURSION TO THE MOON

Weber had started with a resonant-mass detector of a size that could fit in his laboratory, but he had already realized that, in principle, a resonant-mass detector could be much larger. One idea along this line is to consider using heavenly bodies such as the earth itself, the moon, or other planetary objects as resonant masses that would ever so slightly be deformed by passing gravitational waves. A planet has a natural frequency of oscillation, just as the resonant zylinders have, but at a much lower frequency. If it would be possible to measure oscillations of planets at low frequencies, one could indeed try to measure gravitational waves this way.

In 1972 Weber did indeed succeed in attempting such a mission. He led an effort to deploy a gravimeter instrument on the surface of the moon that would measure slight changes in gravity. As the surface of the moon would move up and down from oscillations caused by gravitational waves, the local gravity would appear to change.

The gravimeter instrument, that was sensitive to acceleration in the vertical direction, was deployed on the surface of the moon by the last Apollo mission crew, Apollo-17, on December 12, 1972. Figure 2.3 shows the gravimeter on the moon.

Figure 2.3 The lunar gravimeter, deployed by the Apollo-17 crew in December 1972. The instrument can be seen in the foreground of the image with the upper structure serving as a sun shade. Further back is the electrical and communication support station for the gravimeter and for other experiments at this site. (Image courtesy of NASA, U.S. National Archives.)

After installation, it turned out that the gravimeter could not be precisely adjusted to work in the reduced gravity on the moon, compared to earth. The Apollo crew could make the instrument work with a mechanical tweak; however, the sensitivity of the instrument at low frequencies was compromised from this operation. The data was still useful for seismic studies, but clearly the sensitivity to gravitational waves exciting the moon was less than it could have been. There were other problems with the temperature control of the gravimeter, which made it unavailable for some time, and eventually the instrument and other instruments placed on the Apollo-17 mission was shut down in 1975. A final report from 1977 stated that gravitational waves had not been found.

In recent times, missions to the moon are discussed more frequently and there may be future prospects for similar attempts. After all, such a 'detector' would operate in a frequency band that is not easily accessible to earth-based laser interferometers (see Chapter 7). Besides the moon, also the Mars moons, Phobos and Deimos, have been proposed as possible targets for such gravitational-wave missions. While in principle the earth could also serve as a resonant mass to detect gravitationl waves, the seismic activity on earth makes such attempts more difficult than on the much quieter moon.

2.5 THE FURTHER DEVELOPMENT OF RESONANT ANTENNAS

In the 1970s, although predictions varied considerably as to the observable rate of various astrophysical events likely to be a source of gravitational waves, scientists agreed that detectors would probably need to be up to a million times more sensitive than Weber's cylinders in order to measure gravitational waves with some reasonable probability. In order for resonant antennas to achieve a considerably higher sensitivity, the dominant noise source, called thermal noise, had to be reduced. This was accomplished by operating the detectors at very low temperatures, i.e., near the absolute zero point of the temperature scale.

Subsequently, cryogenic projects with resonant antennas began in the United States (Stanford and Baton Rouge), Australia (Perth) and Italy (Frascati and Legnaro) and at CERN. The introduction of cryogenic technology indicates a new level of complexity in the further development of detectors. In addition to the low temperature requirement, this new cooling technology had to be subjected to seismic isolation, because the cooling techniques produced vibrations and glitches that would affect the sensitivity of the detector. Cryogenic temperatures require a multi-layered design of temperature shields and the insulation of the detectors against seismic influences became more complex and multi-staged. Liquid helium was used to cool down to temperatures of a few Kelvin. A noteworthy achievement is that two of these new projects reached ultra-cryogenic temperatures that are well below one Kelvin using dilution refrigerators.

To convert the vibrations excited in the cryogenic cylinders into electrical signals, new and increasingly improved sensors were required. The result was the development of SQUIDs (Superconducting Quantum Interference Devices). These are extremely sensitive devices that use quantum effects to measure magnetic fields. Using suitable transformers, the mechanical vibration of the cylinder is transmitted into small magnetic field fluctuations which are then measured with a SQUID.

It is important to note that the extreme level of complexity inherent to these detectors can merely be outlined here. Progress came slowly because of the difficulty of the task at hand. In 1986, the first joint data run occurred, using three cryogenic detectors (Stanford, Baton Rouge and CERN). The following year, improvements were being made to the cryogenic detectors such that none of them was in operation when Supernova 1987A was observed on February 24th in the Large Magellanic Cloud, a dwarf satellite galaxy of the Milky Way. A supernova had not happened this close to earth since 1604, and although it is unlikely that the cryogenic detectors could have identified gravitational waves from this event, it is perceived as highly unfortunate to have missed such a rare event.

By the year 2000, five cryogenic resonance antennas were in operation. In 2006, the Italian science agency, INFN, discontinued funding for the research and development of resonant antennas; for several years, however, the three European detectors Auriga, Explorer, and Nautilus continued to operate (see Table 2.1).

Table 2.1 Cryogenic resonance antennas. All detector cylinders are about 3 meters long, are made of an aluminum alloy (with the exception of Niobe) and have a mass of about 2.3 tons. Niobe's cylinder consists of 1.5 tons of the element niobium, the largest object ever made from this chemical element. The operating time covers the period from the first data recording to the end of the project.

Project	Location	Temp.	Frequency	In operation
Allegro	Baton Rouge	4 K	ca. 900 Hz	1991-2007
Auriga	Legnaro	0.1 K	ca. 900 Hz	1997-2016
Explorer	CERN	2 K	ca. 900 Hz	1990-2012
Nautilus	Frascati	<0.1 K	ca. 900 Hz	1995-2016
Niobe	Perth	5 K	ca. 700 Hz	1993-2001

The construction of these cylinders was a remarkable technical achievement. The cryogenic cylinder of the Nautilus detector (see Figure 2.4), for example, was the coldest massive object in the known world. Although these cryogenic cylinders were up to one hundred thousand times more sensitive than Weber's, no gravitational waves were measured. Instead, some other research results were obtained with the cylinders, such as an answer to the question as to whether or not very cold, massive objects behave as predicted by quantum physics. (They do!)

Figure 2.4 The Nautilus detector in an open condition. The resonant bar can be seen in the center of the image, surrounded by several layers of shielding for the cryogenic operation. The image illustrates the complexity and the development the bar detectors have taken over the course of decades. (Image credit: ROG group.)

In addition to the cylindrical resonance antennas, there were also projects in development that intended to use spherical resonance antennas, although most remained mired in the planning phase. The advantage of a spherical shape versus that of a cylinder is that a sphere is susceptible to gravitational waves from all directions, while a cylinder preferably oscillates in the direction of its longitudinal axis and can only receive waves from directions that excite this particular vibration. Nevertheless, even a spherical detector is only sensitive to gravitational waves in a relatively narrow frequency band. This fact was to be regarded as a decisive disadvantage to the use of resonant antennas when compared to interferometers. The spherical detectors could only be built in the form of small prototypes, and interferometers slowly took over the field.

CHAPTER 3

Michelson's legacy: the interferometer

In his quest to detect gravitational waves, Weber used bodies of matter that would be excited by the waves. In 1956, due to Pirani's theoretical work which investigated how gravitational waves interact with matter, it became more clear that it may be possible to measure the change in the distance between freely moving objects that would be caused by a gravitational wave. This created an alternative means for detecting gravitational waves using light.

3.1 WAVES, INTERFERENCE AND THE INTERFEROMETER

A wave can be described as the propagation of an excitation, often carried by a medium. A gravitational wave will curve space and spread in the medium of space-time. A wave spreads both spatially and temporally, which means that at a fixed location, a wave causes an object or a physical quantity to oscillate around a resting point. For example, in a water wave the water molecules on the surface of a lake oscillate and in a light wave electric and magnetic fields oscillate. With spatial extension, the mountains and valleys of the wave form a pattern that changes and propagates over time. The superposition, or the sum of the patterns of different waves, leads to the phenomenon called *interference*. Interference is merely the result of the addition of two or more waves.

Waves and interference phenomena play a role in various areas of physics such as in acoustics and optics. With Young's double-slit experiment of 1801, optical interference phenomena notably led to the description of light as a wave phenomenon — one can think of the colored iridescence of a soap bubble caused by optical interference. Figure 3.1 A shows two waves whose mountains and valleys occur at precisely the same locations and are, therefore, in-phase. For the sake of clarity, the waves are drawn spatially separated. The two in-phase waves in Figure 3.1 A, on the left side, interfere

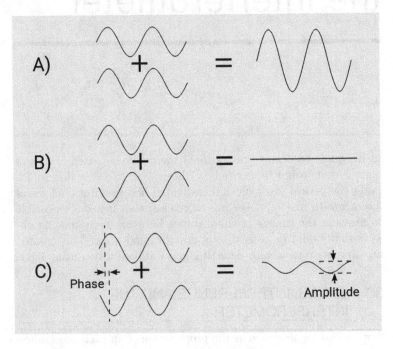

Figure 3.1 The interference of waves.

to form the wave on the right side, which has twice the amplitude of the individual waves. This case is called *constructive interference*. Figure 3.1 B describes the case of *destructive interference* in which the waves on the left side oscillate with opposite phase. This means that the wave crests and troughs lie in opposition, so that they cancel each other out and result in a wave of amplitude zero. Figure 3.1 C shows a small difference in the phase of two waves. In this case, the waves add up to a wave with an amplitude that is proportional to the phase difference between the two individual waves. This means that a small change in the phase between the individual waves is translated into a small change in the amplitude of the resulting wave. In an arrangement where two waves pass through different paths, they experience different phase changes. If the waves are then superimposed, their phase difference is translated into a measurable amplitude — this arrangement is called an *interferometer*.

An interferometer is an instrument that uses the phenomenon of interference to precisely measure a certain quantity, typically a phase difference between two light beams. Since the phase of the light beams is, among other things, proportional to the distance they have travelled, an interferometer is ideally suited for length measurements.

The American physicist Albert Michelson, who was born in Prussia in 1852, had long been interested in the problem of how to measure the speed of light. During a study visit to Germany, he improved upon the predominant method of using a rotating mirror to measure the speed of light, with his idea of using optical interference. In 1881, in Berlin, he built an interferometer, today known as the Michelson interferometer, that measured the relative speed of light in two perpendicular beam paths. At the time, the experiment was used to test the hypothesis that light travels in a medium called ether. He proposed that the motion of the earth relative to the hypothetical ether would result in a delay in light propagation time and, thus, to a phase shift of the light in one arm of the interferometer. Michelson quickly discovered an important source of disturbance in the interferometer operation — ground vibrations. As previously discussed, this issue was also encountered in the operation of resonant antennas. Since ground vibrations hindered the measurement, Michelson moved the sensitive instrument to a quieter location, the town of Potsdam at the outskirts of Berlin. Figure 3.2 shows a photograph of a reproduction of Michelson's early interferometer in the historic location in Potsdam.

In 1887, together with the chemist Edward Morley, Michelson repeated the experiment in an improved form in Cleveland, Ohio. This famous Michelson-Morley experiment provided perhaps the most important negative result in the history of physics — the ether-medium could not be found. Light propagates at the same speed in all directions, regardless of the movement of the light source or the movement of the measuring instrument. This finding is one of the cornerstones of the Special Theory of Relativity. It is not entirely clear, however, whether Michelson and Morley's experimental result played a role in Einstein's development of the Special

Figure 3.2 A reconstruction of Michelson's early interferometer in the historic location in Potsdam, Germany. (Image credit: Dierck E. Liebscher.)

Theory of Relativity, but some sources suggest that Einstein was aware of their result before the development of the theory.

Overall, the course of this story is fascinating. Michelson and Morley used an instrument in their experiment that resulted in a negative result which pointed in the direction of Special Relativity and, thus, to General Relativity. And almost 130 years after their results, gravitational waves, a consequence of the General Theory of Relativity, have been detected with a highly developed Michelson interferometer!

Figure 3.3 shows a schematic drawing of a Michelson interferometer. A light source (a laser, in this case) sends a beam to the beam splitter, a semi-permeable mirror. The light is split into beams b and c and sent to mirrors X and Y. These two paths are called the arms of the interferometer. The mirrors (at the arms' ends) reflect beams d and e back to the beam splitter and they are split into beams f and g, and h and i, respectively. (Here, the beams are drawn spatially separated, but, in the real interferometer they are exactly superimposed onto each other so that they will interfere.) Since beam h passes through one arm and beam f passes through the other arm, small changes in the path length of the arms are translated into small changes in the phase between these two beams. This phase change causes an amplitude change in the interfering beam, i.e., the sum of beams h and f and also g and i. The phase change between beams h and f is registered with

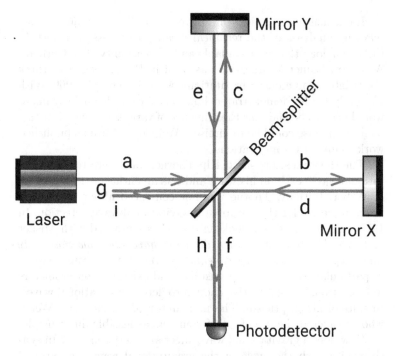

Figure 3.3 Scheme of a Michelson interferometer.

a photodetector as a change in brightness at the output of the interferometer. Beams g and i move in the direction of the laser. Their phase change is typically not registered, but the beams will be important for a reason explained further below.

3.2 THE MICHELSON INTERFEROMETER AS GRAVITATIONAL WAVE DETECTOR

The idea of using a Michelson interferometer to measure gravitational waves appears to have surfaced among various scientists independently of one another. This makes it difficult to determine to whom the credit is due. According to Joseph Weber's lab notes, this idea came to him soon after the Chapel Hill Conference in 1957, and he spoke of it in a telephone conversation with his colleague Robert Forward in September, 1964. The idea is also mentioned in a 1962 Russian publication by Mikhail Gertsenshtein and Vladislav Pustovoit.

Implementing the idea in a real experiment required far-reaching analyses of the technical means, a process tied both to· the technology that was available and its feasibility. The German-American Rainer Weiss, who was born in Berlin, began to think about interferometers as gravitational wave detectors in 1969. With his analyses, he demonstrated how sensitive such an instrument would need to be and how the influence of various sources of disturbance and noise could be minimised. Weiss cites Pirani's published work as one of his inspirations.

The theoretical physicist, Kip Thorne, was interested in gravitational waves early in his work and was an enthusiastic supporter of Weber. Initially, Thorne was not convinced about developing interferometers for the purpose of gravitational-wave detection. In 1970, in a standard textbook on gravity co-authored with Misner and Wheeler, he writes: *Such detectors have such low sensitivity that they are of little experimental interest.* At the time, lasers, in particular, were still very unstable and existing interferometers were far from being sensitive enough to detect gravitational waves. In spite of his skepticism, Thorne maintained contact with Weiss, who considered the use of interferometers as feasible, in principle.

How does a gravitational wave affect an interferometer? In synchronicity with the cycle of the gravitational wave, one arm of the interferometer becomes slightly longer, while the other arm becomes slightly shorter. As previously explained, this difference causes a small, but clearly discernible, phase shift of the light waves in the two arms relative to each other. Figure 3.4 shows a strongly exaggerated deformation of a Michelson interferometer by a gravitational wave. The five sub-pictures demonstrate a full cycle of the passage of a gravitational wave. These times exemplify a gravitational wave with an oscillation frequency of 100 Hz, i.e., 100 oscillations per second.

The Michelson-Morley interferometer had a sensitivity of about 10^{-10} for relative ˙path length differences in the arms. In order to successfully measure gravitational waves with some probability, their sensitivity had to be a trillion times higher, requiring a resolution to relative length changes of 10^{-22} over a timescale of 0.01 seconds.

There are three essential design factors involved in improving the interferometer sensitivity: (1) the length of the arms (the longer, the better, with a restriction explained below); (2) the amount of internal noise made by the component parts, i.e., the

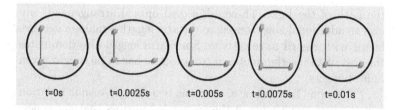

t=0s   t=0.0025s   t=0.005s   t=0.0075s   t=0.01s

Figure 3.4 Compression and stretching of a Michelson interferometer by a gravitational wave. The wave comes from a direction perpendicular to the image plane. Gravitational waves stretch and compress the space perpendicular to their direction of propagation, so they are so-called transverse waves. There are are two possible polarizations of gravitational waves, called plus (+) and cross (×). The diagram shows a plus polarized wave. In a cross polarized wave, the elliptical shape of the stretching of space is rotated by 45 degrees compared to the plus polarization.

mirrors (the quieter, the better); (3) the amount of light (the more light, the better, with a restriction referred to in Chapter 7).

The first factor, arm length, is significant because the longer the arms of the interferometer, the stronger the gravitational wave signal. This is true if the travel time of the light in the arms is shorter than half the oscillation duration of the gravitational wave to be measured. A wave with an oscillation frequency of 100 Hz has an oscillation period of 10 milliseconds. This means that in half of this time, while one arm of the interferometer is stretched by the gravitational wave and the other is compressed, light covers a distance of 1500 kilometers (932 miles). Therefore, the optimal arm length for a Michelson interferometer to measure this gravitational wave frequency would be 750 kilometers (466 miles), since the light travels back and forth the length of the arm. Interferometer arms this long would be very difficult or impossible to realize on earth, so instead clever optical arrangements are used to effectively extend the path of the light. These additional optical arrangements are also an additional noise source, so the arm length remains a decisive factor with regard to sensitivity. Since arm length also dominates the cost of an interferometer, a compromise has to be sought given finite budgets.

The second factor, noise, concerns two things: seismic isolation and the tiny movements of the suspended mirrors that are generated by thermal fluctuations in the materials that constitute them. According to the so-called fluctuation dissipation theorem, friction losses of the mirrors and their suspensions are directly connected with the undesired movements of these components. The theorem says that if there is dissipation (friction) there is also fluctuation (motion). A small amount of internal friction is always unavoidable due to the way the mirrors are suspended. In order to minimize such disturbances, it is important to use low-loss materials with low internal friction, and construction methods that avoid any additional friction. This is the reason why in the highly sensitive gravitational-wave detectors, the *test masses*, that are the most sensitive mirrors, are suspended with glass fibers instead of the more common steel wires: glass has lower internal friction than steel.

The third factor, light, is important because more light increases the measurement accuracy of the phase difference between the beams of the interferometer arms (as a reminder, the phase difference is translated into a measurable light power). Light has been described as a wave in order to explain the phenomenon of interference. It must be added here that light interacts with matter in a quantized form. Individual 'packets' of light, the photons, as mentioned in Chapter 1, are emitted and detected, a discovery which goes back to Max Planck and Albert Einstein. Normally, the time intervals between photon detections are random, resulting in a certain level of noise in the detection of the interfered light. This noise is called shot noise, a term derived from the seemingly random impact of the bullets from a shotgun. Shot noise is an important limitation to the sensitivity of optical interferometers. If more light is used, the disturbing shot noise increases because there are more photons, but the signal strength of the phase difference measurement increases even more, so more light helps!

3.2.1 Let there be light!

The development of the laser enabled decisive progress in interferometry given the requirement for the strongest possible light source. Because of their unique design, lasers produce very high intensity light, emitted as a bundled beam in one direction. In addition, laser light is monochromatic, which means it has a very well-defined wavelength (equivalent to the color of the light), making it perfect for use in interferometers. Simply put, the more monochromatic the light, the easier it is to accurately measure its phase. There are technical limits to the light power output of a laser, mainly determined by the type of laser and its stage of development. In addition to power, the stability of the wavelength and the amplitude of the light are decisive factors in determining their suitability for use in gravitational wave detectors. Control systems are used to stabilize these variables and will be discussed further in Section 3.2.5, "Control is necessary!".

The lasers that are currently used in interferometric gravitational wave detectors are called diode-pumped solid-state lasers. They have an output power of 25 to 200 Watts and a light wavelength of 1.064 micrometers, which is about one thousandth of a millimeter (1 inch=25.4 mm). This is in the near infrared range of light and is invisible to the human eye, making it a bit more impractical in comparison to working with visible light. Previously, gas lasers had been used that emitted visible light, but they were far less stable, such that in the 1990s the switch was made to the new technology.

In addition to the use of the most powerful lasers, specific arrangements of the interferometer mirrors (optical configurations) are used to increase the light power. The current configuration of the gravitational wave detectors Advanced LIGO and Advanced Virgo is described below.

3.2.2 The Fabry-Perot resonator

The Fabry-Perot resonator was developed in 1897 by the French physicists Charles Fabry and Alfred Pérot. It consists of two partially translucent mirrors that are arranged parallel to one another at a precisely tuned distance. This arrangement allows for constructive interference in the space between the mirrors for particular wavelengths of light, where the length of the resonator corresponds to an exact multiple of the wavelength. Fabry and Pérot originally developed the device to separate different wavelengths of light from each other, but the constructive interference also increases the power of the light that circulates in the resonator by adding the light field of many round-trips inside. This increase of light power does not violate the conservation of energy, but is based on the accumulation, or storage, of light.

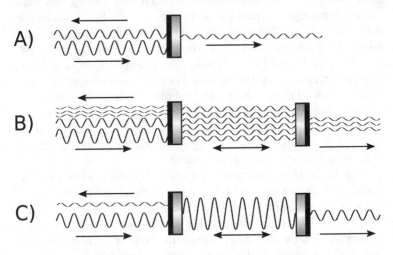

Figure 3.5 Principle of a Fabry-Perot resonator.

Figure 3.5 illustrates how a Fabry-Perot resonator works. The partial illustration Figure 3.5 A shows only one mirror onto which a light wave is incident from the left. The mirror (in this example) is constructed in such a way that it transmits about one third of the light wave and reflects about two thirds. Based on this situation, in Figure 3.5 B, a second mirror is added, which together with the first mirror, forms a Fabry-Perot resonator. The waves in the resonator between the two mirrors are now reflected back and forth several times. With each reflection, a part of the wave is transmitted through the respective mirror. On the left side of the resonator, the waves coming from inside have the opposite phase of the directly reflected wave, which results in partially destructive interference. Figure 3.5 C shows the individual waves of Figure 3.5 B, now summed up as a single wave. In the resonator space between the mirrors, the constructive interference of all partial waves results in a light wave with a larger amplitude than that of the wave incident from the left side, an effect known as resonant enhancement.

Just how much larger the resulting light wave gets is dependent upon how much light both mirrors let pass.

In a Michelson interferometer optimized for the detection of gravitational waves, Fabry-Perot resonators have two important functions. First, up to four such resonator arrangements are used to increase the amount of light in the interferometer, to extend the effective arm length, and to optimally extract a gravitational wave signal. Second, Fabry-Perot resonators are used as so-called *mode cleaners*, optically cleaning the light rays, or, more precisely, geometrically and temporally filtering them. This use of a Fabry-Perot resonator as a mode cleaner is not further explained here, but is an example of the technical complexity of gravitational-wave interferometers.

3.2.3 Fabry-Perot resonators expand the Michelson interferometer

Using the concepts developed in the Fabry-Perot resonator, how can the light output be increased in a Michelson interferometer? Figure 3.3 demonstrates that the interferometer also sends light back towards the laser (beams g and i). The amount of light sent to the photodetector and the amount of light sent back to the laser depend upon the phase difference of the light waves in the arms. The person operating the interferometer will choose a specific point, an operating point, during the undisturbed state when there is no gravitational wave to affect the phase difference of the light in the two arms. A potential gravitational wave signal then generates small fluctuations around this operating point. The operating point can be adjusted so that most of the light is reflected back to the laser and very little light reaches the photodetector. The sensitivity of the interferometer does not depend on this operating point but on the amount of light present in the arms, even though this may come as a surprise and is sometimes presented differently.

If an operating point is selected that allows almost all of the light to be sent back to the laser, the Michelson interferometer now acts as a mirror. Given this fact, another mirror can now be inserted between the laser and the beam splitter of the interferometer, so that a Fabry-Perot resonator is created that allows the amount of light circulating in the arms to be resonantly increased, and thus also makes the interferometer more sensitive. This technique is called *power recycling* because the light traveling from the interferometer back to the laser is not discarded and unused, but is sent back towards the interferometer by the mirror called a *power recycling mirror* — a brilliant idea!

The tales of the German "Schildbürgers", or *The Fools of Schilda*, come to mind. The builders of the local town hall had forgotten to install windows! The Schildbürgers tried to collect sunlight in bags and various containers in order to illuminate the interior. Unlike this vain attempt, the use of power recycling in today's gravitational wave interferometers actually allows light to be stored for about one second — not too bad at all!

Another idea to increase the sensitivity was to use Fabry-Perot resonators *in the arms* of the interferometer. This allows light to travel back and forth in the arms multiple times, increasing the phase shift of the light in the arms generated by a gravitational wave. This concept, however, has a price — an increased phase shift, and thus increased sensitivity to gravitational waves, only occurs from waves at low frequencies, the longer gravitational waves. With shorter gravitational waves, the travel time of the light in the interferometer arms is longer than the period of the gravitational waves, such that some of the signal is lost again. For this reason, a design compromise must be made that takes into account the frequency distribution of the expected gravitational waves and various technical boundary conditions. In combination with the power recycling concept, using Fabry-Perot resonators in the arms further increases light power within the interferometer. In today's large interferometers, the light power provided by the laser is resonantly increased by the power recycling technique and the Fabry-Perot resonators in the arms by a factor of about 5000. This is only made possible by the use of very low-loss mirrors that reflect up to 99.999 % of the light.

Fabry-Perot resonators are only one possible way to extend the effective light path in the arms. Prior to the development of Fabry-Perot resonators, another idea was pursued for this purpose — delay lines. With a delay line, the light is reflected back and forth in each arm several times before the light from both arms is superimposed on the beam splitter. In contrast to the Fabry-Perot resonator, in a delay line the beams remain spatially separated. Delay lines require larger mirrors or the use of several mirrors in order to achieve multiple reflections. One distinct advantage to their use is that the process of locking the resonators is not necessary. (Locking will be discussed in the upcoming Section 3.2.5.)

In addition to power recycling and placing resonators in the arms, another optical resonator constructed with yet an additional mirror can be formed at the output of the interferometer, between the beam splitter and the photodetector. The result is called *signal recycling*. The process of signal recycling is somewhat subtle, but it can be said that it optimizes the extraction of the gravitational wave signal from the Fabry-Perot resonators in the arms. If there are no resonators in the arms, signal recycling will amplify the signal, having a similar effect as the arm cavities. In addition, signal recycling allows the interferometer to be operated in a so-called narrow-band mode. This means that the sensitivity is greater in one range of frequencies than in others. For example, in order to make the detector as sensitive as possible to certain astrophysical sources, like colliding neutron stars, the frequency of the best sensitivity of the detector could be selected in order to improve the chances or the accuracy of detection. Figure 3.6 shows a Michelson interferometer with power recycling, signal recycling and Fabry-Perot resonators in the arms.

3.2.4 Seismic isolation

Michelson and Weber, as previously noted, took great care to isolate their detectors as much as possible from environmental influences, particularly from ground vibrations. To an even greater extent, the mirrors of a modern interferometric gravitational wave detector must be shielded from environmental influences, especially the mirrors that mark the ends of the arms, the test masses. Ground movements must be suppressed by factors of more than one billion so as not to disturb the distance between the test masses, which is read out with the highest of precision. In addition to the optical reading of the phase difference between the arms, seismic isolation of the test masses is another complex task that exemplifies the degree of physical science and engineering required to construct highly sensitive interferometers.

There are two principles employed in seismic isolation. The first is to measure and actively suppress undesired ground movements. The second is to arrange the most sensitive mirrors in the form of cascaded pendulums, a passive isolation technique. We will only be considering the second principle.

Let's imagine a button tied to a piece of thread to make a pendulum. A pendulum has an interesting property in that movements of the upper end of the thread are transmitted to the button attached to the bottom of the thread in a certain form, depending on the frequency of excitation. If the upper end of the thread is

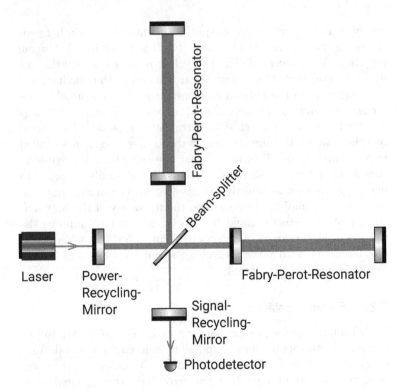

Figure 3.6 A Michelson interferometer enhanced by Fabry-Perot resonators. The width of the lines representing light fields between the mirrors symbolizes the different power levels of the light in the different segments.

moved very slowly sideways, the button follows this movement almost without delay. However, if the movement of the upper end of the thread occurs very quickly, the button will follow with a delay, due to its physical inertia. In this way, movements of the upper end of the thread at high frequencies, i.e., with fast movements, are transmitted to the button in an attenuated form. This results in an isolation effect at these frequencies. The motion of the button is being isolated, at least to a degree, from the motion that governs the motion of the thread. The gravitational wave detectors currently in operation employ this principle of isolation by engaging up to seven cascaded pendulum stages to support the test masses.

It is interesting to note that between very slow and very fast movements, there is also a frequency at which the movement of the upper end of the thread leads to resonantly enhanced movement of the object at the bottom. This effect is not desirable within the seismic isolation system and is actively suppressed by technical means in the interferometer.

3.2.5 Control is necessary!

Control loops are used to suppress undesired oscillations in the pendulum stages. In a control loop, an actual value of a system is measured, compared to its desired value, and an actuator is engaged to exert an influence on the system in order to bring about the desired value. A classic example of a control loop is a heating thermostat: the thermostat measures the room temperature, compares it with the desired temperature and increases or decreases the supply of heat to the room, thereby adjusting the room temperature to the desired value. In the suspended pendulum stages, the movements of the mirrors are measured and forces are exerted with electromagnets in order to keep the mirrors as still as possible.

In an interferometric gravitational wave detector, there are also much more complex control loops that continuously maintain the operating point of the Michelson interferometer, as well as the precise length at which desired constructive interference is produced in the Fabry-Perot resonators. One important art of interferometer design is to develop systems that provide precise information about all relevant mirror positions in order to control them with feedback loops during operation. To ensure that the control process itself does not cause too much disturbing noise, all measuring systems and control loops must be precisely designed to be compatible with the ultimate sensitivity desired from the gravitational wave interferometer.

The process of putting these many control loops into operation is not trivial. The teams working on the various interferometer projects had to invest a considerable amount of time in this task, which is also referred to as the *lock acquisition* or *locking* of the interferometer. Locking describes the process of snapping the length of the many resonators to a multiple of the light wavelength in order to reach the resonance condition. The reason why this is difficult is that many of the signals used for the control of the mirror positions are only available once the optical resonaters are locked, such that this constitutes a chicken and egg problem. One way to solve this is to engage additional signals for the mirror positions which are less accurate, but help to bring all mirrors close to the desired positions. Once the long sequence of steps required to achieve the locked state has been found, an attempt is made to automate the process with the aid of computer programs.

In order for the gravitational wave interferometer to record significant measurement data, it must be in the locked state. However, this state can be temporarily lost, again and again. In particular this can happen when the control circuits do not have sufficient force to counteract an external event, such as ground vibrations caused by a remote earthquake. The interferometer must then be brought back into the locked state and this takes some time.

Today, almost all control loops in interferometric gravitational wave detectors are implemented in digital form. The sensing system measures the positions of the mirrors, information which then gets digitized. Computers calculate the necessary control signals in real time and give out control signals to actuate on the mirrors. The control parameters can be changed while the system is running and this flexibility has proven indispensable for the optimization and commissioning (see Chapter 4) of interferometers.

3.2.6 Vacuum system

In addition to the laser, the optical design, the mirror isolation stages and the control circuits, it is important to note that the majority of the interferometer components are located in an ultra-high vacuum, where the usual ambient air pressure is reduced by a hundred billion times! Such a vacuum system is necessary for two reasons. First, it serves to keep acoustic disturbances away from the optical components and, most importantly, from the test masses; otherwise, the vibrations of the air molecules in the sound waves will cause the test masses to move too much. Second, the greatly reduced pressure ensures that not too many air molecules pass through the light beam in the arms of the interferometer. This is important because every molecule that passes through will lead to a tiny, undesired phase shift in the light beam.

3.2.7 Does a gravitational wave not stretch the light as well?

Before concluding this section, we may briefly look at a worthwhile question that is sometimes brought up when thinking about whether light can indeed be used to detect gravitational waves in the way described above. One can ask the question: Does a gravitational wave not squeeze and stretch a light beam as well? And if it does, does this not mean that light cannot be used to measure gravitational waves, since it would be squeezed and stretched in the same way as space?

Somewhat surprizingly the answer to the first question is yes: A gravitational wave does indeed squeeze and stretch a light beam, that is any beam that is already present in the space between the mirrors when a gravitational wave strikes. However, the answer to the second question is 'no', and that is because what we measure with the interferometer is not the wavelength of the light, but its travel time, which expresses itself as a phase difference. So in the very first moment a gravitational wave strikes the interferometer, nothing is detected, since the waves of the light are stretched in space. However, the laser will continue to feed new light into the detector, such that after a short while there will be phase difference detectable in the two beams returning from the two arms. The wavelength of these new beams is stretched no longer by the gravitational wave from a moment ago, and thus now we get a signal, just with a small delay.

3.3   THE PROTOTYPE INTERFEROMETERS

We now turn back to history — the development of prototype interferometers. A prototype interferometer is used to hone the techniques required to operate a gravitational wave detector. From the outset, the chance of an actual detection with a prototype, however, is negligible. The probability of detection depends on both the sensitivity of the interferometer and the frequency and strength of possible events that may generate gravitational waves.

Robert Forward from Hughes Aircraft Research Laboratory in Malibu, California, and a former member of Joseph Weber's team, was the first scientist to begin building an interferometer as a prototype in 1971. With a simply folded arm length of 4.25 meters (about 14 feet), this instrument achieved about the same sensitivity to gravitational waves as Weber's cylinder. It had an advantage over cylinders though, which were only sensitive in a narrow band around 1660 Hz, in that it was sensitive in a broad frequency band. However, the further development of this instrument was discontinued and Forward turned his attention to another area of study.

Beginning in 1972, at the Massachusetts Institute of Technology (MIT) in Cambridge, MA, Rainer Weiss attempted to obtain research funding from the National Science Foundation (NSF). It was finally granted in 1975, but, initially, Weiss had difficulties getting PhD students to work on his project because it involved lengthy development work. At that time, the cylinder antennas had been established and the future of interferometers was still uncertain. In 1975, Weiss said in an interview (Collins, 2004): *We (at MIT) are in a physics department. And ... engineering is not considered respectable physics. To build something and show that it works as predicted, but without making a measurement of anything new does not really count as any achievement.* Despite this obstacle, Weiss started with a prototype arm length of 1.5 meters (about 5 feet) and, in 1981, was able to secure funds for a study to build a much larger detector with arm lengths in the kilometer range (1 kilometer=0.62 miles).

In Munich, in 1974, Heinz Billing's group turned from cylinders to interferometers and began to build a laboratory-sized prototype with an arm length of 3 meters (about 10 feet). Using the concept of delay lines, beams passed through each arm of the interferometer up to 138 times. This was the world's leading interferometer for many years and served the development and successful demonstration of important new interferometer techniques: To avoid disturbing mechanical resonances, the hanging of mirrors as pendulums was realized by Karl Maischberger; the invention of the mode cleaner, by Albrecht Rüdiger and others, suppressed disturbing laser beam movements; and Walter Winkler developed a comprehensive theory of the effect of scattered light. At about the same time, the concept of power recycling was proposed by Roland Schilling in Munich and by Ronald Drever in Glasgow. Figure 3.7 shows an image of the 3-meter prototype in Munich.

In 1983, the construction of a much larger and improved prototype, with an arm length of 30 meters (about 99 feet), began on the Garching science campus near Munich. This prototype was the first of its kind in the world to reach shot noise, an important limitation to the sensitivity of optical interferometers that had previously been only theoretical. This achievement was to be of decisive importance in the funding of the American LIGO project.

By the end of the 1980s, the sensitivity of the Garching detector was about 10^{-19}. In comparison to Weber's cylinders, twenty years earlier, this is an advance of a factor of 1000, in addition to having a much wider frequency range. The 30-meter prototype was in use until 2002 and in its final years it was the first interferometer to demonstrate the combination of power and signal recycling, an

Figure 3.7 The 3-meter prototype in Munich, Germany, showing the individual light beams of the delay line, made visible using artificial fog. (Image: Max Planck Society, Photographer: Peter Blachian.)

optical configuration known as dual recycling. It also served as a test facility for the GEO 600 detector in Germany, prompting the development of such techniques as the ability to keep the suspended mirrors at the correct angle during a measurement.

In Glasgow, Scotland, beginning in 1975, Ronald Drever also turned his attention to interferometry, initially studying it in order to achieve a more precise readout from resonant mass antennas. In 1976, the construction of a prototype interferometer with an arm length of 10 meters (about 33 feet) began, applying the concept of Fabry-Perot resonators in the arms. In 1979, following an invitation from Kip Thorne, Drever also led the construction of a 40-meter (about 131 feet) arm-length prototype at the California Institute of Technology (Caltech). The Caltech prototype was later used to develop locking technology for Advanced LIGO, among other things. After Drever moved to Caltech permanently in 1983, Jim Hough took over the management of the 10-meter prototype in Glasgow. Subsequently, the Glasgow Group developed the concept of signal recycling (Brian Meers), which was later implemented on the 30-meter prototype in Garching and on the GEO 600 detector.

In addition to the first significant prototypes that we have discussed, another noteworthy facility is the Australian International Gravitational Observatory (AIGO), located north of Perth. Originally, the construction of an interferometer with arms several kilometers long was planned here, but, despite concerted effort, the necessary funds could not be obtained. AIGO is currently a prototype with an arm length of 80 meters (about 263 feet). It is used for testing high intensity light power in interferometers.

The degree of difficulty inherent to this field of work is illustrated by Heinz Billing in a quote written in 1977 in the journal *Physik in unserer Zeit* (translated by the author): "*The developers of the 'new' gravitational wave antennas have undoubtedly set themselves a very interesting but also very difficult target. Ongoing tests will show whether it can be achieved in a few years' time. Whether it will be achieved, however, depends not least on the willingness to provide the necessary, certainly not inconsiderable, means for such a profound experiment, which at first glance only satisfies the human thirst for knowledge.*"

CHAPTER 4

Interferometers around the world

In the mid-1980s, after gaining experience with laser interferometer prototypes and developing new techniques, groups in the United Sates, the United Kingdom and Germany, and later in France and Italy, began applying for research funding for kilometer-sized systems (1 kilometer=0.62 miles). This was a tall order given that at least a hundred million dollars would be needed to build instruments that would have at least a small chance of ever measuring gravitational waves. Unlike accelerator technology in particle physics, such as the Large Hadron Collider (LHC) at CERN, interferometer technology was not yet a mature science and considerable uncertainty remained as to whether or not such large facilities would function at all, and, if they did, how well they would function. There was considerable risk involved in the whole endeavor!

4.1 LIGO

The LIGO project (Laser Interferometry Gravitational Wave Observatory) was initiated in 1984 by Rainer Weiss and Kip Thorne, with the help of other scientists, and was supported by both the California and Massachusetts Institutes of Technology, Caltech and MIT. Initially, LIGO was jointly managed by Drever, Thorne and Weiss but their collaboration lasted only three years, as all decisions had to be made by consensus and Drever and Weiss often had differing viewpoints on technical issues. In 1987, progress was accelerated when the National Science Foundation (NSF) urged the appointment of a single director, Rochus Vogt, to lead the LIGO project. An example of Vogt's contribution was the executive decision to use Fabry-Perot resonators to increase the effective path length of the arms, as proposed by Drever, and not to use the delay line technique favored by Weiss.

In 1989, LIGO scientists submitted an application for funding to the NSF, proposing facilities with four-kilometer-long arms (almost 2.5 miles) at two locations in the United States. The infrastructure, consisting of buildings and vacuum systems, was envisioned to be used for multiple generations of interferometers. Due to the uncertainty of interferometer technology at this newly proposed larger scale, the scientists suggested that the first interferometer to be installed, LIGO I, later renamed to Initial LIGO, would remain technically rather conservative in order to improve the chances of success. Among other things, it was decided to hang the mirrors (the test masses) for Initial LIGO in simple single pendulum stages, although the groups in Europe were already working on the development of multiple pendulum stages. Given this choice, it became clear that Initial LIGO had only a small chance of measuring gravitational waves and a plan was initiated to replace Initial LIGO with a more technically advanced interferometer, called Advanced LIGO, using the same infrastructure.

The discussions about financing Initial LIGO dragged on for quite a while. There was considerable opposition, particularly from astronomers who feared they would lose the money allocated for their projects, in addition to being irritated by the fact that LIGO described itself as an observatory. For an astronomer, an observatory should indeed observe something, but as mentioned, it was rather unlikely that in the first development stage of LIGO gravitational waves would be detected. In the end, the required funds were pledged in 1992, with the US Congress providing additional support, reducing the likelihood that the National Science Foundation would have to withdraw its support from other projects in order to finance LIGO.

In 1994, Barry Barish and project manager Gary Sanders took over the management of LIGO. Each had extensive experience with large projects in particle physics and understood the need for strict management. The leap from prototype interferometers to facilities that were approximately a hundred times larger represented the shift from lab-oriented research to the actual operation of large-scale facilities and required the scientists involved to relinquish part of their control to the managers. During the time of Barish's leadership, the LIGO Scientific Collaboration (LSC) was founded in 1997, fostering national and international collaboration. Among other things, the expertise of other groups was necessary for the successful development of the Advanced LIGO interferometer.

In 1992, the NSF selected Hanford, Washington and Livingston, Louisiana, as sites for the LIGO detectors. From a purely scientific standpoint, Livingston is not necessarily an ideal location, as the soil is damp and, on average, it has more ground motion than other places that were under consideration. Unsurprisingly, it is not only scientific but also political factors which play a role in projects of this magnitude. Politicians, in order to advance their own agendas, will sometimes work very hard to bring a large, high profile, science project to their state.

Construction of LIGO's buildings and vacuum systems began in 1994. In order to construct two, four-kilometer-long, stainless steel tubes (at almost 2.5 miles, they are the largest ultra-high vacuum system in the world), the company CB&I (Chicago Bridge & Iron) built factory-like plants in both Hanford and Livingston. Steel was welded from rolls to tubes, 1.2 meters in diameter (almost 4 feet). This impressive engineering achievement was successfully completed in 1998. Figure 4.1 shows an aerial view of the LIGO site in Louisiana in 2015. The first interferometers, Initial LIGO, which were installed in the infrastructures, included a Michelson interferometer with Fabry-Perot resonators in the arms and power recycling, using mirrors suspended in single pendulum stages, as previously mentioned. In addition to the two interferometers built in Hanford and Livingston, each with an arm length of 4 kilometers, Hanford also received an interferometer with a 2-kilometer arm length (almost 1.25 miles) in the same vacuum system. After an installation phase of almost three years, the 2-kilometer interferometer in Hanford was the first LIGO interferometer brought into a locked state in 2001, which took several months to accomplish.

4.1.1 Commissioning

A large interferometer is a complex system consisting of many subsystems, the most important having already been mentioned: laser and optical configuration, seismic isolation systems, control systems and the vacuum system. Commissioning, in a general sense, describes the process by which a technical system is prepared and adjusted. This involves searching for any potential errors and formulating solutions to eliminate them. To a certain extent, all

Figure 4.1 LIGO in Livingston, Louisiana. (Image courtesy of Caltech/MIT/LIGO Laboratory.)

subsystems of the interferometer are subject to this process when they are first set up. Commissioning, in a narrow sense, includes the investigation and optimization of all subsystems and their interaction with each other while the interferometer is in operation, since some undesirable properties may only be discovered during simultaneous operation of all subsystems.

The first step in commissioning is the work of bringing the interferometer into the locked state (see Chapter 3). The basic problem to be addressed involves the signals necessary to keep all mirrors in their desired positions using the control loops. As mentioned before, these signals only exist in a very small area around the optimal mirror position and at the beginning of this process one is largely blind to the current, rather random, mirror positions; therefore, special methods must be developed to provide the signals needed for locking. For example, additional lasers can be installed to obtain signals for the mirror positions over a large range. These signals are not good enough to read a gravitational wave signal, but do serve to reach the locked state.

Once the locked state can be reliably reached again and again, the work of analyzing and improving the sensitivity of the interferometer begins. This describes the second step in commissioning. Since the path length differences of the arms to be measured are extremely small, the measurement is extremely sensitive and the entire system is susceptible to various types of disturbance and noise. A typical commissioning cycle consists of measuring the sensitivity of the interferometer and analyzing which subsystems limit the sensitivity at a given frequency. This work often resembles detective work in that various hypotheses are formulated and tested. If the cause of a limitation in sensitivity can be identified, there is an attempt to eliminate it or, at least, reduce it.

For example, during the commissioning of Initial LIGO, fluctuations in the frequency of an oscillator used to generate the signals required to control the mirrors were causing a limitation in sensitivity. Fluctuations of the oscillator frequency can disturb the length signal of the interferometer in various ways. In principle, these couplings can be calculated, but the information required for these calculations is not always available, in addition to the fact that new coupling paths are sometimes discovered that had not been foreseen. In this case, in order to reduce the noise contribution, the original oscillator was replaced by a more stable one, specifically designed for this purpose.

Another, perhaps more mundane, example is that even cables can be a source of noise, if they have been damaged by missing strain relief, or if they are not of sufficient quality for long-term and large-sacle installations. Lessons like these had to be learned by the teams of all interferometer facilities.

Once a noise source is eliminated or reduced, the cycle of analysis and improvement continues until the interferometer reaches its target sensitivity. The target sensitivity is mostly limited by noise sources which can be described as fundamental, that are inherent to the design of the interferometer and its subsystems. This means that further improvements in sensitivity can only be achieved by either making significant changes to subsystems or completely replacing the interferometer.

After several years of commissioning, Initial LIGO achieved its target sensitivity in 2005. Between 2002 and 2010, a total of six data runs were performed, which lasted from a few weeks to more than two years. The aim of a data run is to keep the interferometer in a state that is sensitive to gravitational waves for as long as possible. For the most part, commissioning does not take place during this time, in an attempt to make as few changes as possible. It may take place to a very limited extent if absolutely necessary.

The purpose of this is to achieve as much consistency as possible in the characteristics of the collected data.

In 2010, as the data runs continued, the detectors' sensitivity was partially increased in commissioning phases and reached a value of about 2×10^{-23} at 200 Hz. Since this statement reflects an increase at only one frequency (200 Hz), another measure of sensitivity is employed that takes into account a larger portion of the frequency spectrum. This is achieved by using the calculated gravitational wave signal of two merging neutron stars (see Chapter 5) and estimating the distance to which such a source can be detected by an interferometer. In 2010, LIGO achieved an average range of about 20 megaparsec (Mpc) for two merging neutron stars, which means it could detect this type of source up to a distance of about 66 million light years. One megaparsec corresponds to a distance of about 3.3 million light years.

4.1.2 Advanced LIGO

Research and development work on Advanced LIGO, the interferometer that was to replace Initial LIGO, began around 2004. It involved Australian scientists, as well as those from the UK and Germany, who were part of the GEO collaboration.

When compared to Initial LIGO, Advanced LIGO demonstrated several technical improvements: a stronger and highly stable laser with a light output of 200 watts (a contribution from the Max Planck Institute for Gravitational Physics in Hannover, Germany), active seismic pre-isolation of the pendulum suspensions, quadruple pendulums for seismic isolation of the test masses, larger and heavier test masses (40 kg, or 88 lbs) suspended on glass fibers, electrostatic actuators and signal recycling. With the exception of the active pre-isolation, these techniques were initially all developed in the GEO collaboration in Great Britain and Germany, and partially tested in the GEO 600 detector (discussed in the upcoming section, "GEO").

Financing was finally secured in 2008 and the Advanced LIGO project received $200 million for three interferometers — one in each of both Hanford and Livingston, within the existing infrastructure, and a third interferometer also to be built at Hanford. The plan for the third interferometer was later changed with the intention of setting it up in another country due to the benefit of source localization (see Chapter 5). Initially, Australia was interested in providing an infrastructure for this interferometer but decided in 2011 to participate in the Square Kilometre Array radio telescope project. This meant that funding for a large interferometer would not be possible, given the needs of the radio telescope project. In the end, it was India that was to become the third location of an Advanced LIGO detector, called LIGO-India.

Under the project management of David Shoemaker, the construction of the two Advanced LIGO interferometers in Hanford and Livingston took about four years and was completed in 2014. This marked the start of the commissioning phase for these instruments. The experience that had been gained with the first generation of interferometers allowed for rapid progress in improving sensitivity above frequencies of about 100 Hertz. However, progress was slower in the frequency range below 100 Hertz, a region that was now possible to explore for the first time, given the new quadruple pendulum suspensions of Advanced LIGO. The task was to eliminate other sources of disturbance at these lower frequencies, such as, for example, unwanted electrical charges on the test masses that interact with electromagnetic fields in the environment. Another problem at low frequencies is scattered light, mentioned in the previous chapter, which can be generated by a tiny amount of unevenness or dust contamination on the test mass surfaces. If this vagabonding light finds its way back into the main beam of the interferometer, it inserts a small, disturbing phase shift. The radiation pressure of scattered light can also push the test masses ever so slightly, an effect that, in turn, can disturb the measurement of gravitational waves.

Overall, Advanced LIGO's objective was to improve sensitivity by a factor of 10 compared to Initial LIGO. About one third of this improvement (an increase in sensitivity by a factor of about 3) was achieved by both Advanced LIGO detectors by the beginning of their first data run in September 2015. This corresponded to a range of about 60 Mps, or 200 million light years, for the detection of binary systems of merging neutron stars.

4.2 VIRGO

Since the late 1970s, the French physicist Alain Brillet had been interested in the detection of gravitational waves and visited Rainer Weiss at MIT in 1980 and 1981. Adalberto Giazotto worked at the University of Pisa in the mid-1980s on multi-stage seismic isolation systems with the intention of using them for gravitational wave detectors. In 1985, Brillet and Giazotto met at a conference in Rome and decided to work together with the aim of building a large interferometer. The potential collaboration with the German interferometry group in Garching did not unfold because the management of the German group assumed that their own project would soon be approved and a collaboration could cause delays. For this reason, Brillet and Giazotto started their own project — Virgo. This name choice expressed the hope that the Virgo interferometer could be used to measure gravitational waves of astronomical objects from the Virgo cluster. This is the closest large cluster of galaxies to Earth, at a distance of fifty million light years.

Initially, the French scientific organisation CNRS had not provided any funds for this project. However, in 1989, two other Italian groups, Frascati and Naples, joined Virgo. A new application was submitted both to the CNRS and the Italian scientific organisation, INFN. Both agencies finally approved the project in 1993 and 1994, respectively. Cascina, near Pisa in Tuscany, was chosen as the location for the Virgo interferometer.

Just as in the case of Livingston, Louisiana, Virgo's location is not ideal, as interests besides scientific considerations were at work in the determination of the site location. It took some time to acquire the necessary land due to the fact that it was owned by over fifty individuals. In addition, after completion, the main building of the Virgo interferometer proved to be susceptible to water ingress and the soft ground caused the interferometer arms to slowly sink, not unlike the phenomenon affecting the Leaning Tower of Pisa. Virgo uses hydraulic presses distributed along the entire length of the interferometer arms to counteract the slow sinking.

Many scientists working on the Virgo project came from particle physics and thus had experience with accelerators, but not with interferometry. In spite of that, in order to increase the likelihood of faster progress, it was decided not to build a prototype but to develop the necessary techniques using Virgo. The Virgo project has a less hierarchical structure than the LIGO project and is based more on a collaborative model. Many individual university groups were involved and this fact makes strict project management much more difficult. From 1996 to 1999, there wasn't any unified project management and this delayed Virgo's construction. This led to the foundation of the EGO (European Gravitational Observatory) consortium in 2000, which was tasked with designing, operating and planning improvements to the interferometer.

Like LIGO, Virgo chose the optical configuration using Fabry-Perot resonators in the arms of the Michelson interferometer and the technique of power recycling to increase the light power. However, from the very beginning, the scientists relied on Giazotto's proposed, ambitious seismic isolation system for the test masses, consisting of seven cascaded pendulums that together formed an eight-meter-high tower. This design, known as Super Attenuator, would presumably improve sensitivity at low frequencies more than any project thus far (down to around 10 Hertz), even in the interferometer's first stage of development. The project did not progress very quickly because of this decision to take such a major technological step. Despite all the handicaps, the interferometer was completed in 2003 and the commissioning work began. Joint data runs with LIGO took place from 2007 to 2010. In 2011, after one last data run together with the GEO interferometer, the first generation Virgo interferometer was switched off to make room for Advanced Virgo.

In addition to institutes in the founding countries of Italy and France, the Virgo collaboration, as of 2017, included institutes in the Netherlands (joining in 2006), Poland and Hungary (joining in 2010), and Spain (joining in 2016).

4.2.1 Advanced Virgo

Advanced Virgo is the name of a new second-generation interferometer that aims to improve sensitivity about tenfold compared to the first generation of detectors, just as in the case of LIGO. The Advanced Virgo project was approved at the end of 2009, almost two years after Advanced LIGO, with a significantly smaller budget of around 20 million euros. Among other things, this meant that a somewhat riskier optical configuration had to be chosen, theoretically making it more difficult to operate the interferometer with high light power.

Advanced Virgo encompasses a similar list of improvements as Advanced LIGO, but the seismic isolation by the super attenuators remains essentially unchanged. Virgo does not use electrostatic actuators to apply forces to the test masses. Instead, small magnets are glued to the test masses and force is applied with electromagnets. In the first expansion stage, Advanced Virgo runs without signal recycling, which reduces the complexity of the operation and allows for the locked state to be reached more easily. It was in late 2016 that the Advanced Virgo interferometer was first brought into the locked state. In the summer of 2017, after a commissioning phase, Advanced Virgo participated in the final phase of Advanced LIGO's second data run (O2). Even with the short commissioning time that was available, Advanced Virgo's sensitivity was already improved in comparison to the first generation of Virgo. In August 2017, a range of 26 Mpc (78 million light years) was achieved for gravitational waves from binary systems of merging neutron stars.

To give an impression of the size and complexity of the optics, Figure 4.2 shows an image of the Advanced Virgo beam splitter.

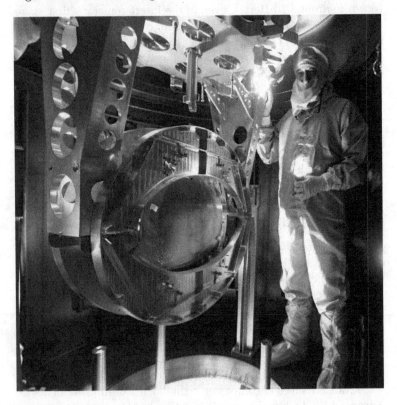

Figure 4.2 Advanced Virgo beam splitter at the lower end of a Super Attenuator. The beam splitter in the middle of the image is framed by a metal screen, which absorbs scattered light. Since all optical components must be kept extremely clean in the vacuum chambers, the scientist on the right in the picture wears a cleanroom suit. (Image courtesy of the Virgo Collaboration, Photographer: Maurizio Perciballi.)

4.3 GEO

Beginning in 1985, inspired by the progress and findings made on the prototypes in Munich and Garching, the German gravitational wave researchers tried to obtain funds for a detector with an arm length of 3 kilometers (almost 2 miles). However, the German research funding organizations were not sufficiently interested in the project. In 1986, a similar situation occurred in the UK, where the application for funding to build a large interferometer submitted by Jim Hough's group was also rejected.

In the following years, these two groups merged to form the GEO Collaboration and a German-British research proposal was submitted in 1989. In addition to experimental centres in Garching (Germany) and Glasgow (Scotland), groups from Cardiff University (Wales), University of Strathclyde (Glasgow), University of Hannover (Germany), University of Oxford (UK) and Physikalisch-Technische Bundesanstalt Braunschweig (Germany) participated. This application was submitted to the then Federal Ministry of Research and Technology (BMFT) in Germany and the British Science and Engineering Research Council (SERC). It proposed the construction of either an interferometer with an arm length of 3 kilometers (almost 2 miles) in the Harz mountains (Germany) or a facility with an arm length of 2.6 kilometers in Scotland. The costs were estimated at an equivalent of about 50 million dollars. Despite positive evaluation, the project was ultimately rejected by both the BMFT and the SERC. In the aftermath of German reunification, in 1990, GEO was not on the top of the list of German science policy priorities.

After this disappointment, Karsten Danzmann, who had taken over the management of the Max Planck Group in Garching at the end of 1989, came up with the plan to build a much smaller, and therefore cheaper, facility. Thanks to ambitious technology, the German-British GEO 600 detector was anticipated to be competitive with the larger detectors in at least part of the frequency band. After the University of Hannover and the State of Lower Saxony provided suitable land in Ruthe near Sarstedt, south of Hannover, Germany, the GEO 600 detector with an arm length of 600 meters (0.37 miles) was created. The German portion of the financing was contributed by the Max Planck Society, the Volkswagen Foundation and the State of Lower Saxony, and the British portion by the PPARC (Particle Physics and Astronomy Research Council). The construction of GEO 600 began in September 1995, under the direction of Karsten Danzmann who later became the director of the Department of Laser Interferometry and Gravitational Wave Astronomy at the Albert Einstein Institute (a Max-Planck institute) in Hannover. A large portion of the detector's infrastructure was built by the scientists and students themselves. An innovative design of the 600-meter-long vacuum tubes, and a reduction to the bare essentials of the buildings, made it possible to save money on material and infrastructure cost.

The optical configuration of GEO 600 consisted of a Michelson interferometer with dual recycling, the combination of power and signal recycling. Fabry-Perot resonators were not used in the arms because the simultaneous use of arm resonators and dual recycling had not yet been tested on prototypes. Initial LIGO did not use this combination for the same reason. Instead of arm resonators, GEO 600 used simple folded arms (delay lines with just one additional mirror in each arm) to double the effective path length. From 2003 to 2009, GEO 600 was operated in a moderately narrowband mode of signal recycling, which included the development of new techniques for mirror angle adjustments and for achieving the locked state.

The ambitious technology of GEO 600 was also an opportunity to test this technology for possible use in the larger detectors in the future. From the beginning, GEO 600 used triple pendulum suspensions and, since 2003, utilized glass fibers for the suspension of all test masses in the lowest pendulum stages. In comparison to conventional steel wires, glass fibers result in lower mechanical friction, which reduces thermal noise. In addition, electrostatic actuators, which do not require that magnets be glued to the test masses, were the method chosen to exert forces on the test masses. The routine operation of these innovations at GEO 600 bolstered confidence in these techniques. The triple pendulum suspension, including glass fibers and electrostatic actuators, were then further developed by the GEO collaboration into Advanced LIGO's four-stage suspension of test masses.

The sensitivity of GEO 600 did not reach that of LIGO, however, mainly due to the fact that LIGO's arms are seven times longer. The specific design choices of GEO 600 also brought a challenge the other interferometers did not face to the same extent: Without Fabry-Perot resonators in the arms, all the light power must pass through the beam splitter. Inevitably, a small proportion of the light is absorbed in the beam splitter, heating it up and causing a lens effect. The returning rays from both arms thus have slightly different shapes and can no longer be brought to destructive interference effectively. As a result, some unused light leaves the interferometer at the output, giving rise to a variety of problems: power recycling may not work as well, there is more stray light in the interferometer, and the sensor systems needed to control the mirrors can be disturbed. Such a thermal effect by absorption of light, as in the beam splitter of GEO 600, ultimately represents a technical limitation with regard to light power in all interferometers. With the help of innovative techniques, such as the targeted heating of certain areas of the beam splitter or the test masses, the scientists managed to reduce this problem in GEO 600.

From 2002 to 2010, GEO 600 participated in joint data runs with LIGO and Virgo. Unlike LIGO and Virgo, no second generation interferometer was planned for GEO 600. Instead, an incremental upgrade was implemented, starting in 2008. Once again, new techniques would be tested and, from 2010 to 2015, GEO 600 was the only interferometer recording observational data for long periods of time. This minimized the risk of missing a possible cosmic event of great strength while LIGO and Virgo were shut down for the installation of the second interferometer generation. In the period from 2010 to 2018, GEO 600 recorded observational data, on average, two thirds of the time. A high degree of automation allowed for such an intensive measurement operation and also made it possible for it to be carried out by a small team.

In 2010, GEO 600 had its most important upgrade — the application of squeezed vacuum, a technique that reduces shot noise. Shot noise (Chapter 3) can be interpreted differently with a modern view from quantum theory in which there is no state of an absolute void of energy. This leads to the so-called vacuum fluctuations to which every electromagnetic field is subject. When a light beam is detected, the electromagnetic vacuum fluctuations interfere with the electromagnetic field of the light wave, giving rise to shot noise. In order to lower the shot noise, however, optical techniques can be used to reduce the vacuum fluctuations, a condition known as squeezed vacuum.

As early as 1981, the use of the squeezed vacuum technique to improve gravitational wave detectors was proposed by the American theoretical physicist Carlton Caves, but it took decades of laboratory work by groups in Germany, Australia and the US to make his idea a reality. While this new technology was briefly tested on a first-generation LIGO detector in 2011, it has been in permanent use at GEO 600 since 2010 and is being continually improved.

The squeezed vacuum technology has recently, in 2018, been implemented in Advanced LIGO and Virgo and is being used in the third observational run O3 (See Chapter 6 for O3).

Figure 4.3 shows a view into the central building of GEO 600.

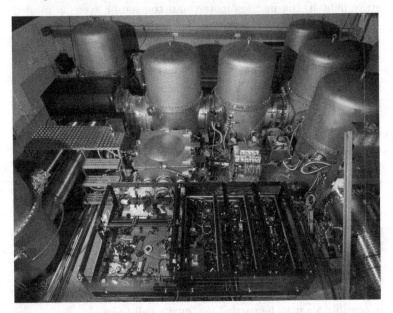

Figure 4.3 A view into the central experimental building of GEO 600. Qn the table in the foreground, we see the structure that creates the squeezed vacuum. In the background, we see some of the central vacuum vessels of GEO 600 where the mirrors and test masses are suspended as pendulums. (Image courtesy of the Max Planck Society.)

4.4 KAGRA

Among the countries that entered experimental gravitational-wave work early on was Japan. TAMA 300 was an interferometer with 300-meter long arms (328 yards) located in the city of Mitaka near the Japanese capital, Tokyo, on the campus of the National Astronomical Observatory of Japan (NAOJ). It was operated by the Institute for Cosmic Ray Research (ICRR). TAMA 300 was the world's first interferometer with arm lengths of over 100 meters for measuring gravitational waves. The optical configuration used Fabry-Perot resonators in the arms, as well as power recycling.

The construction of TAMA 300 began in 1995 and the interferometer was first locked in 1998, initially without power recycling. The first data recording took place in 1999 and, in 2000, a new sensitivity world record of 10^{-21} was set — TAMA300 was more than ten thousand times more sensitive than Weber's cylinder. In 2002, a joint data run with LIGO was accomplished.

The seismic environment in the Tokyo area is strongly influenced by ground vibrations caused by human activities. For this reason, the seismic isolation systems were upgraded to multiple pendulums in 2005. In 2011, however, some of the components of TAMA were damaged in the strong Tohoku earthquake and regular operation as a sensitive interferometer was subsequently discontinued. The infrastructure continued to be used, however, for technology development.

Another interferometer prototype, called CLIO, was built in Japan at the Kamioka Mine in Gifu Prefecture. The Kamioka Mine is known by physicists for its neutrino (subatomic particle) detector, Super-Kamiokande. CLIO, with its 100-meter-long arms (328 feet), was mainly used to test the effectiveness of the technique used to cool the interferometer test masses down to 20 Kelvin (−253°C or −424°F). Cooling reduces the thermal noise of both the test masses and the lowest pendulum stage. The Japanese collaboration planned to use this technique for their larger detector.

Early plans for this large detector in Japan included two identical interferometers, each with three-kilometer arms (almost 2 miles), at one location. After considerable delays, a single interferometer was finally approved in 2010. The project was named KAGRA — KA for Kamioka and GRA for gravity. KAGRA was built into the Kamioka Mine complex and the interferometer is located completely underground. In this environment, ground vibrations in the frequency range of 1 to 10 Hertz are typically one hundred times less than in the Tokyo area and ten times less than in a quiet, above-ground location on earth. The seismically calm environment helps to control the mirrors and makes it easier to increase the sensitivity of the interferometer at low frequencies. The cryogenic operation of the test masses requires a different test-mass material than the amorphous quartz used for the room-temperature interferometers: KAGRA is using test masses made of sapphire.

The earthquake in 2011 and the resulting impact on the Japanese economy delayed the start of construction and there were further setbacks during the excavation of the two, three-kilometer-long tunnels for the arms due to severe water inrushes. In spite of this, the construction of the tunnels was successfully completed in 2014 and, by the end of 2015, the vacuum tubes for the interferometer arms had already been installed. Figure 4.4 shows a view of one of the vacuum tubes along an arm of the interferometer.

Figure 4.4 The y-arm tunnel of KAGRA, one of two 3-kilometer long underground beamlines. The vacuum tube containing the laser beam can be seen extending from the left side into the mountain. (Image courtesy of NAOJ.)

Unique to the Japanese research funding system is the separation of human and material resources. This meant that the core team for setting up the interferometer was initially quite small given the magnitude of the project. Nevertheless, under the leadership of Takaaki Kajita (who was awarded the Nobel Prize in Physics 2015 for the discovery of neutrino oscillations), KAGRA made rapid progress in establishing the infrastructure and interferometer components.

An initial interferometer consisting of single pendulum suspensions in a simple Michelson configuration was sucessfully brought to a brief operation phase in 2016. The installation of all optics for the full cryogenic interferometer with mutiple-pendulum suspensions was nearly completed in early 2019 and commissioning of parts of the full interferometer had begun. The KAGRA detector is expected to join the O3 run, possibly later in 2019, and contribute to sky localisation and parameter estimation of gravitational-wave events. In addition to its astrophysical importance, the operation of KAGRA will also be valuable to gain experience with underground and cryogenic operation of a large interferometer. Both of these techniques are expected to be used for future gravitational-wave detectors on earth (see Chapter 7).

CHAPTER 5

Data analysis and Big Dog

To build gravitational wave detectors that were sufficiently sensitive to actually detect signals required hundreds of scientists and engineers dedicated to research and development spanning decades. The detectors record a vast amount of data and in order to uncover the waves the data must be analyzed. Since this task is so specific and labor intensive, a specific field of gravitational wave research emerged, called *data analysis*. It is mainly carried out by four research teams, each specializing in one type of signal (with certain overlaps).

In principle, data analysis can distinguish between (a) modelled and unmodelled searches, and (b) between transient (short-term) waveforms and quasi-continuous ones. If these two criteria are arranged in a matrix, four forms of data analysis are obtained, as shown in Table 5.1.

Table 5.1 Four forms of data analysis and the corresponding gravitational-wave sources and waveform types.

	Modeled	Unmodeled
Transient	merging binary systems	burst, impulse
Quasi-continuous	rotating neutron stars	stochastic background

A modeled search is guided by a theory that predicts a waveform for a particular type of source. Such waveforms can be calculated for several source types and they describe the time-sequence of space deformation by a gravitational wave. Scientists then search for this exact waveform in the data collected from the detector, while also looking for variants in order to cover different possible parameters of the gravitational wave source.

In contrast, the unmodeled search is not based on any particular signal form, but merely looks at deviations from the normally expected noise in the output signal from one detector and compares them both with noise deviations from other detectors (coincidence concept) and with signals from environmental sensors (veto concept). The modeled search is specific and sensitive; the unmodeled search is more general and less sensitive. Joseph Weber carried out an unmodeled search in the data of the resonant antennas, where the veto concept and the coincidence concept (see

Chapter 2) helped him to distinguish possible gravitational waves from statistical or disturbing fluctuations.

The classification of a signal as transient is subject to a certain arbitrariness, since, in practice, no signal can be strictly permanent. When searching for gravitational waves, waveforms shorter than a few seconds are typically classified as transient. In contrast, quasi-continuous signals are usually not subject to time limitation. The prefix 'quasi' reminds us that there are no truly permanent processes in nature, even if a signal is regarded as stationary and unchangeable for a given period of time within the scope of the data evaluation.

5.1 MODELED SEARCH: MATCHED FILTERING

Matched filtering describes a method of searching for a known, or expected, waveform in a data stream. Mathematically, it can be demonstrated that the search for an expected signal is most effective, i.e., technically delivers the optimum signal-to-noise ratio, if this signal is precisely known.

When using the matched filtering method, the values of a calculated waveform at successive time steps are multiplied with the values of the detector data at successive time steps and all resulting products are totaled. This process is repeated for each time step, synchronous with the way the data has been recorded, and provides a new data stream. If the desired signal is present in the new data stream, then, depending on the size of the signal in the detector data, the matched filtering method will show this as a signal that can be distinguished from random noise fluctuations. For this signal, as in the case of the unmodeled search, threshold values must be defined; if they are exceeded, the signal found is subjected to further analysis.

The matched filtering method can only find waveforms that were previously derived from theory. Because it is possible that waveforms may be present in the data that were not previously predicted by any theories, an unmodeled search is also necessary in order to cover the entire spectrum of possible waveforms.

5.1.1 Waveforms of merging binary systems

Two orbiting compact astronomical objects, usually neutron stars or black holes, are called binary system or compact binary system (see Chapter 2). Due to the radiation of gravitational waves, a binary system loses energy and the distance between the two objects is reduced; this process inevitably ends with the merging of the objects.

The process can be divided into three phases. The first is the inspiral phase: Initially, the compact objects circle around each other, slowly approaching. Due to the radiation of gravitational waves, their circling speed gets faster and faster. The second is the merger phase: The objects make contact and merge into one another. The third is the post-merger or ring-down phase: The new object, created during the merging, reduces any remaining asymmetry by emitting further gravitational waves.

In the case of two merging neutron stars, a calculation for the gravitational waves to be expected in each of these three phases was made around the turn of the millennium. Certain characteristics of the waveforms depend upon the properties of neutron star matter and these are not yet very well known. Therefore, repeated measurement of the gravitational waves of merging neutron stars will provide scientists with completely new insights into this extreme form of matter.

The waveform resulting from the fusion of two black holes was a harder nut to crack. It was regarded as one of the most difficult mathematical-physical problems to be solved and several research teams around the world were working on the solution. At this time, also around the turn of the millenium, it was uncertain as to whether such signals would possibly be picked up by the detectors before they even could be calculated!

While the first and third phases of the black hole merger process could be reasonably estimated by approximation methods, Einstein's equations of general relativity had to be solved numerically, i.e., using computers, for the merger phase. Once again, the non-linearity of these equations became a stumbling block and the computer programs were initially unable to calculate the process from start to finish. In 2005, the problem of creating a complete simulation of the fusion of two black holes was solved, almost single-handedly, by the South African Frans Pretorius. Pretorius, who worked at Caltech and the University of Alberta, adopted various approaches by his colleagues and combined them in a new way in order to solve this problem. Shortly thereafter, other groups of scientists using different methods made the same breakthrough and were able to confirm the waveforms calculated by Pretorius.

Today, catalogues with several hundred thousand templates are used. Templates are a designation for each calculated waveform and cover varying masses of the objects involved in the binary system, from approximately one to several hundred solar masses (one solar mass is the mass of the sun). Templates can be short or long, depending on the signal duration in the sensitive band of the interferometer. Scientists routinely search for systems consisting of two neutron stars, two black holes or a neutron star and a black hole. The search for binary black hole systems with much larger masses is not carried out routinely yet, because these systems would emit signals at frequencies too low to be reliably detected with earthbound interferometers.

5.1.2 Rotating neutron stars

In principle, the search for the gravitational waves of rotating neutron stars employs the matched filtering method, albeit in a slightly different form. A neutron star with mass asymmetry emits gravitational waves quasi-continuously (see Chapter 1). The waveform produced corresponds in close approximation to the waveform produced by a sinusoidal oscillation, which can be used as a template for the matched filter analysis. To maximize the signal-to-noise ratio of the matched filter, the templates are as long as possible — in this case, months. This is also referred to as integration, which means collecting and accumulating the signal over a long period of time.

Due to the rotation of the earth around its own axis and its orbit around the sun, an important modification must be made to the waveform used for matched filtering. From the point of view of a gravitational wave detector on earth, these two movements result in a so-called Doppler shift of the received signal. When the detector moves toward the gravitational wave source, the received frequency increases slightly. When it moves away from the source, the detected frequency is lower. For each direction in the sky, the Doppler shift causes a unique pattern of modification to the signal so that many different modified templates are required to analyze the data of long-lasting signals, such as those from quasi-continuous sources. Given the multiple potential combinations of these effects, the search for continuous gravitational waves emitted by rotating neutron stars is particularly complex and requires a high level of computing capacity. With more computing power, larger sectors of the sky can be searched for neutron stars rotating at different speeds. In order to increase the level of computing power, this type of search has been integrated into a program that makes use of computing capacity made available by volunteers, particularly private users. Einstein@Home uses a volunteer computer's idle time to search for signals from spinning neutron stars. While gravitational waves of spinning neutron stars have not yet been found (as of summer 2019), the Einstein@Home program is also used to find new pulsars in the data from radio telescopes. Volunteers have already discovered about sixty neutron stars with this search!

In cases where the celestial position and rotational speed of a neutron star can be ascertained from radio astronomical observations (see pulsars, Chapter 7), a very specific search for their gravitational waves can be performed. In this case, because it is an identified neutron star with known attributes, this kind of search has fewer unknowns than when searching for previously unidentified neutron stars and their gravitational waves. This means that more computing time can be invested in the time integration of the possible signal, such that gravitational waves from known rotating neutron stars can be detected to further distances. The Cancer and Vela pulsars are examples of neutron stars whose sky positions and rotation parameters are well known. For both of these pulsars dedicated searches for quasi-continuous gravitational waves have been carried out. While none have been found yet, the energy that can be emitted in the form of gravitational waves by these objects has been constrained by the observations.

To further complicate the data analysis, the rotational speed of neutron stars can change slowly or in sudden spurts. A slow change is caused by the star's continuous energy loss. A small part of this energy loss is caused by the radiation of its gravitational waves where some of the rotational energy of the star is converted into wave energy. Potential changes in rotation speed must also be taken into account when using templates. If the rotational speed changes suddenly (called a glitch), additional modified templates must be used for the time periods before and after the glitch. In addition, such a glitch event can cause the emission of a short-lived gravitational wave pulse, also called a burst.

5.2 DATA ANALYSIS IN A DETECTOR NETWORK

The search for waveforms using matched filtering can be performed on data from a single detector. However, such a signal would probably only be classified as a gravitational wave if this kind of waveform had been confidently detected several times before by simultaneous observation in more than one detector. If several detectors are operated as a network and their data is evaluated together, there are two major advantages: First, the significance of a detection increases. Second, the direction from which the signal came can be determined.

The slight difference in arrival time at the detectors at different locations provides information about the direction from which the wave originated. This is comparable to what happens when hearing with two ears versus one ear: If an acoustic signal reaches one ear earlier than the other, the brain uses this time delay to reconstruct possible directions for the sound source in the room. The shape of the auricles (the visible part of the external ear) provides additional directional information. Evolution has made some animal species true masters in determining the direction of a sound, whether for locating prey or fleeing from predators.

The accuracy of the directional determination increases with both the strength of the received signal and the distance between the detectors. In addition, the orientation of the detectors in space also provides additional information to narrow down the direction of the source. Unlike optical telescopes, an interferometric gravitational wave detector is sensitive to most directions from which a wave might strike, but there are orientations of the detector with respect to the source with reduced sensitivity or almost total insensitivity. This fact can still be very valuable in the analysis of the data when determining directional information, provided that other detectors have picked up a sufficient signal.

Figure 5.1 shows the locations and orientations of the gravitational-wave detectors on earth.

5.2.1 Burst search

The search for unspecific bursts of gravitational waves that are not initially derived from a model is even more dependent on network data analysis than the search for modeled signals. One can only assume to have observed an astrophysical source with some probability if an unknown waveform is registered by at least two detectors. If gravitational wave signals from this source are found in the future, it is a huge incentive for the development of ideas and models to explain them.

The burst search is intensified during time intervals when other astronomical instruments are available for observation. The time

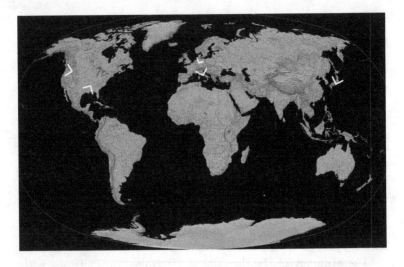

Figure 5.1 Approximate locations and orientations of laser-interferometric gravitational wave detectors on earth. From left to right: LIGO Hanford, LIGO Livingston, GEO 600 (small), Virgo, KAGRA. Detector sizes are not to scale. (Map source: https://ian.macky.net/pat/map/world.html.)

and direction of so-called gamma-ray bursts (GRBs) are used to search for possible gravitational waves that may be connected to these previously mysterious events. For example, merging compact binary systems with at least one neutron star are thought to be a possible cause for gamma-ray bursts. Events in neutrino detectors can also serve as external time stamps. In all of these cases — if another observation has taken place at the same time — the statistical significance of a possible burst signal from gravitational waves is increased.

5.2.2 Stochastic search

A stochastic process is governed by rules of randomness, let it be randomness that reflects insufficient knowledge or fundamental randomness, as postulated by quantum physics. Stochastic gravitational waves are expected from the superposition of a large number of gravitational-wave sources that cannot be individually resolved. In this sense they carry an element of randomness, since the exact amount of space-time distortion cannot be predicted at any time. It is possible though to predict an average level of space-time distortion from stochastic gravitational waves when assuming a certain number of discrete sources throughout the universe; or vice versa, the search for stochastic signals informs models about the abundance of individual sources, such as binary black holes and neutron stars. Two types of stochastic gravitational waves can be distinguished: those from compact binary systems throughout the universe, and those relic from the Big Bang, for example caused by quantum fluctuations in the very early universe, amplified by inflation. The former are sometimes called a *foreground* of stochastic gravitational waves, whereas the latter are called a *background*. The random aspect of stochastic gravitational waves makes them appear as a sort of 'noise', but we treat them as a signal, since we are interested in studying their properties.

As in the burst search, no assumptions are made about the specific waveform in the search for stochastic gravitational waves, which have a seemingly random waveform. In contrast to the burst search, quasi-continuous gravitational wave signals are assumed for the stochastic waves that are hidden in the detector's inherent noise output. By directly correlating the data of at least two detectors, such stochastic gravitational wave signal that is present in different detectors simultaneously can be distinguished from individual detector noise (which is ideally different in each detector).

For the most part, stochastic gravitational waves are expected from all directions in the sky equally, as would be the case for the examples given above. This is the reason why this search works best if the detectors are not too far apart or the signal frequencies are not too high; otherwise, different detectors will not receive the same signal at the same time. Although it would be possible in principle to search for stochastic sources from certain directions in the sky using suitable time shifts between the data of detectors at different locations, there are no processes known that could generate sufficiently large gravitational waves in a preferred direction of the sky.

When searching for stochastic sources, it must be ensured that the detectors do not show any correlated technical noise. Global magnetic field fluctuations are an example of a possible correlated technical noise source. Ideally, these would be picked up by environmental sensors and eliminated during analysis.

5.3 CANDIDATE AND SIGNIFICANCE

To exemplify the data analysis process, the following is a list of concrete steps in the search for transient, or short-lived, signals:

- For each detector, a list of events that exceed a selected threshold is generated. These events are called triggers. They can be unmodeled transient events or events determined by matched filters. The veto concept can be used to remove some triggers from the list.

- The triggers of all available detectors are compared (coincidence concept) and if similar events are registered at almost the same time by different detectors (within the maximum possible signal propagation times), this event receives the status of a candidate.

- To assess the significance of a candidate event, its characteristics are compared with those candidate events (that appear by random fluctuations and disturbances) of detector data in which very likely no gravitational wave signals have been recorded.

So how is the significance of a candidate event estimated? As previously mentioned, Weber had already introduced an estimation, based on detector data, of the number of coincidences between two detectors that are not caused by gravitational waves but by disturbances or random fluctuations. This method is called background estimation (not to be confused with a stochastic background of gravitational waves), which is done using the time-slide method, or time shifts: The data of one detector is shifted by a certain time interval against the data of another detector, allowing the determination of the rate and characteristics of the coincidences that show up during the data analysis. This process is repeated several times with increasingly time-delayed data, for as long as a sufficient amount of data is available or until a sufficiently accurate estimation of the background has been obtained. Determined in this way, the background (one for each specific signal form) will contain information on how often and to what extent coincidences occur between detectors that cannot, as yet, be classified as gravitational waves without further information. In the final analysis, to estimate how significant a gravitational wave candidate might be, the coincidences of the non-time-shifted data streams are then compared to the background. Given a signal of the detected strength (or stronger), an equivalent measure of significance is the determination of how likely it is that it could have resulted from internal disturbances or random fluctuations in the detector data itself.

Such an estimation of the background is not trivial and researchers often find themselves facing a dilemma: If a gravitational wave signal is actually contained in the data, then one would want to remove it from the data when estimating the background; otherwise, the rate of random coincidences with the time-slide method would be overestimated. The problem is that a signal can only be identified with some certainty as a gravitational wave once its significance has been estimated using the background and, only then, can it be removed from the data.

To resolve this dilemma, a two-step procedure is used: First, no triggers are removed from the time-shifted data; second, if clear candidates are obtained, they are removed and the background is determined once again. The first step is key to identifying a candidate for the first time, while the second step increases the significance of this candidate.

The process of estimating the background is similar to carrying out a control experiment, similar to those used in medical studies. For example, one group of patients receives an effective drug, while the control group receives a placebo. Subsequently, the health status of the patients in both groups is monitored and compared. Without a control group, any change in condition would be attributed to the new drug, even if it was actually due to other causes. With a control group, it can be assumed that any differences between the two groups, with regard to the condition being treated, can be attributed to either the medication or the placebo. Any outcomes caused by something other than the medication or the placebo should occur in exactly the same degree in both groups. The control group provides an estimate of the statistical background in this example, precisely because it has not received the actual medication.

Medical and psychological studies are blinded to prevent any influence by the participants on the data collected. This means the participants do not know whether they are receiving a placebo or an effective medication. In the case of double-blind studies, this also applies to those who carry out the study. However, there is still the danger that during the analysis of the data, the conscious or unconscious wishes of the respective scientist or analyst may affect the result. This is where triple-blinding or blind analysis is employed. In this case, the treatment details are also concealed from those who analyze the data. This method is also used in the search for gravitational waves and in some other fields of physics.

5.4 BLIND ANALYSIS

Science is a permanent race between our inventiveness to deceive ourselves and our inventiveness to avoid just this. This quotation, by astronomer Saul Perlmutter, expresses the fundamental danger in the execution and analysis of scientific experiments — namely, the desire to achieve exactly the results we would like. Although Weber used time-delayed data to determine the background as accurately as possible, it can still be assumed that inaccuracies in his analyses, particularly in the handling of statistics, have crept in unconsciously.

Even if one has a method to estimate the statistical significance of a result, when performing the analysis, various settings must be made and the selection criteria defined. For example, a threshold value is chosen and above this value a signal is evaluated as a trigger. In order to avoid the risk that the analyst may choose parameters that will skew the results, i.e., a signal statistically appearing in a more favourable light than is appropriate, the blind analysis method is used.

Blind analysis requires that the data to be analyzed is not viewed until all search parameters have been defined. There are several ways to go about this. In medical studies, the assignment of a drug or a placebo to the participants can be kept secret until the end of the study. When gravitational waves are detected, time-delayed data can be used to search for randomly appearing coincidences. By using this kind of test data, the actual analysis can be run through and tested completely until the analysts agree to remove the blind condition. Only when this last step of the analysis is reached, is the button pressed, so to speak, and the result, in the form of a statistic, is generated.

The usefulness in using this approach in physics is demonstrated by the observation that experimental results are sometimes strikingly close to already existing measured values. For this reason, both particle physicists and cosmologists have introduced blind analysis when evaluating their data.

In the case of gravitational-wave detection, immediately after acquisition of the data by the detectors, the first task is to perform a fully automated evaluation of the data using a computer with the aim of detecting gravitational wave signals as quickly as possible.

Straightaway, other astronomers are informed about a probable detection in order to exploit the full potential of gravitational wave astronomy. Astronomers will search for optical or other electromagnetic signals in the calculated direction of the source using different observational tools. Ideally, this provides information about a potential cosmic event, such as a supernova or the fusion of two neutron stars. This combination of data from varied observational instruments is called multi-messenger astronomy. Multi-messenger astronomy increases the probability of obtaining new insights that would not be possible, if not for this combined effort. For example, the propagation speed of gravitational waves can be estimated more accurately by comparing the gravitational wave signal with electromagnetic signals gathered by other instruments measuring the same source. More on this later!

In addition to the automatic search, a detailed analysis of the data of gravitational wave detectors takes place at a later time when, for example, a planned observational period is completed.

5.5 BIG DOG

On September 16, 2010, the two LIGO detectors register a strong candidate for a gravitational wave signal. This event could have been caused by two merging black holes or by the fusion of a black hole and a neutron star. It appears to have come from the direction of the constellation "Big Dog", hence the catchy name: *Big Dog* event (Figure 5.2).

With the intention of presenting a scientific publication on the Big Dog event, a sequence of steps previously agreed upon within the scientific collaboration is initiated. On March 14, 2011, the work, including a complete manuscript, is completed and about 350 scientists from the gravitational wave research community gather in a hotel near Pasadena, California. Sparkling wine in plastic cups is served before the big reveal; the envelope would finally be opened and the uncertainty wiped away. At this moment, very few members of the collaboration know whether it is a gravitational wave or merely a *blind injection*, a variant of the blind analysis.

In a blind injection, a computer is programmed to feed one or more signals that look like a possibly expected gravitational wave into the detectors at randomly selected, but coordinated times. However, the computer program may also randomly choose not to feed in any signals at all. Within the collaboration, only a select number of researchers know whether or not a blind injection has taken place. Therefore, if a signal that looks like a plausible gravitational wave is found in the data, researchers must also consider that the cause could have been a blind injection.

This approach of using the blind injection protocol was introduced for various reasons. Initially, it served as a test and trial run for an actual detection. Discussions about the significance, the details of the statistical analysis, and the creation of the inevitable scientific publication can be rehearsed before the real event. Another advantage of this procedure for such a large research collaboration is the prevention of leaks to the public. If there would be a leak of information about a candidate event, the public can always be reminded that the event may turn out as a blind injection, and not a real gravitational wave.

In analyzing the Big Dog event, there are many discussions about the problem of estimating the statistical background. If the candidate event is not removed from the time slide data, its significance is slightly less than four sigma, a measure of the statistical significance. According to particle physics standards, this would not be sufficient to meet the criteria of a secure detection; however, it must be noted that there is no fixed rule for the use of words such as detection. In the case of the Big Dog event, removing the candidate event from the time slides results in a significance of five sigma, which would count as a more certain detection. In the planned publication, both statistics would be presented and, in this case, the somewhat more cautious scientists prevailed — the title of the publication refers only to *evidence*, i.e., hints of a gravitational wave, rather than using the more strong word *detection*.

When the envelope is opened by the director of LIGO, it is revealed that the Big Dog was the result of a blind data injection, and not a real gravitational wave! The publication is not submitted and the collaboration must continue to practice patience and perseverance.

Figure 5.2 A star constellation gives rise to an event name and a case of blind analysis unfolds. (Illustration: Josh Field.)

CHAPTER 6

They exist!

In September 2015, after about five years of upgrade work, beginning with Initial LIGO and progressing to Advanced LIGO, the new interferometers are ready for Observational Run 1 (O1), the first round of data acquisition in the advanced detector era. This run was preceded by a so-called engineering run during which time only minor changes to the detectors are permitted, in order to bring the instruments and analysis programs into the most stable state possible. What happens next is a complete surprise!

On September 14, 2015, at 9:50:45 a.m. coordinated Universal Time (UTC), during the engineering run, the two LIGO detectors almost simultaneously register a signal lasting for a few hundredths of a second that corresponds to a maximum relative expansion of space-time of 10^{-21}. In spite of the tiny size of the signal, the disturbance is very clearly visible in the data. Shortly thereafter, the automated search for burst signals registers this event and enters it into a database. Since it is late at night in the United States, a researcher in Hannover, Germany, notified by an automatically generated e-mail, is the first to take note of the new entry in the database. This unleashes a wave of events!

First, the Hannover researchers call the LIGO detector sites in the United States to confirm whether or not something special had actually happened at the time in question. They are told that everything was quiet, and that nothing special was noted. Could it have been another blind injection? Eventually, the entire collaboration is informed of the event, and it quickly becomes apparent that no blind injections were planned during the engineering run! At the time in question, there were also no signs of a targeted injection of signals, which are occasionally used to test the analysis programs.

For the entire scientific collaboration, this event is surprising for at least two reasons: (1) Following a 5-year upgrade break, the detectors had only been recording data that can be meaningfully used to search for gravitational waves for a few days. (2) The signal is unusually clear and can be seen with the naked eye in the data streams; it has a large signal-to-noise ratio. Figure 6.1 shows the data of the two LIGO detectors at the time of the observed signal.

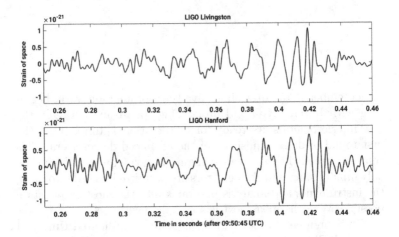

Figure 6.1 Output signals of the two LIGO detectors on September 14, 2005, at 9:50:45 UTC time. The raw data of the interferometers were processed in two steps: They were calibrated to represent them as an expansion (strain) of space, and they were filtered in the range from 35 to 350 Hz (a bandpass filter) to suppress major fluctuations in the detector signal outside this frequency range. Similar signal amplitudes of increasing size can be seen in both time series. The signal shape appears about seven thousandths of a second earlier in the Livingston detector. This event would later be assigned the name *GW150914*, denoting the day it occured.

The waveform roughly corresponds to the expected waveform of two orbiting and eventually merging black holes, which is in accordance with theoretical predictions. In addition, the signals recorded by both detectors are very similar. Together, these facts intuitively convince many researchers in the collaboration that this was indeed the first measurement of a gravitational wave in history! However, at this point, the assessment is not yet strictly quantitative but is based on the signal form visible to the naked eye and its assessment in the given context.

In the following weeks, there is a lot of work to do and a lot to discuss within the collaboration. Establishing scientific findings is by no means self-evident. The efforts to clarify what was observed provide an insight into how knowledge is often processed in the practice of science.

The September 14th event is considered to be a clear candidate for the detection of a gravitational wave. Two days later, the previously agreed upon plan is, once again, put into effect, and ultimately leads to the creation of a publication. A detection committee plays a central role in the evaluation of the candidate. The committee members review the indications and data provided by the various working groups and, if necessary, recommend steps for further analysis. The blind analysis of the event is prepared and the states of the detectors at the time of the event are examined and documented in great detail. In addition, it is decided that the detectors are kept in their current condition, unless changes are urgently needed. The purpose of this directive is to keep the characteristics of the data collected by the detectors as homogeneous as possible in order to provide useful data for the time-slide method, i.e., for the statistical estimation of the background.

Why wasn't this event registered by the automated search for waveforms from merging binary systems? The explanation is readily available: At the time of the event, this form of data analysis only used templates representing binary systems with black hole masses that are lower than those corresponding to the observed waveform. However, the signal was large enough to be detected by the automated search for unmodeled burst signals. In time, the search for signals from binary systems would have employed templates representing binary systems with larger masses as well and, thus, also identified the event.

If it wasn't a blind injection of test signals, could there have been a conspiratorial plot to feed artificial signals into the detectors? Is it possible that computer hackers wanted to play a bad joke on the collaboration? These issues were indeed investigated for a short time, but it was more of an intellectual exercise in preparation for any potential questions along these lines. The conclusion was reached that such a malicious enterprise would have required expertise in several different areas, making it impossible for a single person within the collaboration to pull it off. Several people would have had to join together in a grand plot, making this an extremely unlikely scenario.

Should this event have been considered a serious candidate, even if it occurred during the engineering run before the planned official start of O1? The operators of the detectors explicitly determine the status of the instruments for each time phase. For example, data generated during a commissioning phase is not used to search for gravitational waves. At the time in question, however, all systems in the engineering run were in the status of data acquisition, i.e., intended for data-taking. Therefore, generally, it is not seen as a problem that the event occurred during the engineering run.

Does it seem strange that the event was registered in the first week of data recording with Advanced LIGO? At the start of the engineering run, Advanced LIGO's range was about three times greater than the range of Initial LIGO (see Chapter 4). Specifically, because the volume of the observable universe increases with the cube of the reach of a gravitational wave detector, the number of observable potential sources of gravitational waves is about 27 times greater at this time than for Initial LIGO. The data recorded with Initial LIGO in its most sensitive phase (the s6 data run from 2009 to 2010) was usable for a period of about six months. With three times the range of Initial LIGO, Advanced LIGO has a similarly high chance of detecting an event in just one week. In addition, due to the quadruple pendulum cascades in the suspension of the test masses in Advanced LIGO, there is a significant extension in the detection range for binary systems of black holes (due to the lower frequency signals), so there is nothing strange here!

Even though many members of the collaboration are convinced of the credibility of the signal, a blind statistical analysis is carried out as planned — this alone can certify the significance of the event for experts watching from the sidelines, as well as the wider public. On October 5, 2015, the work to set the parameters of the blind analysis reaches an endpoint. In a collaboration-wide teleconference, data analysts come to the consensus that the analysis can now be conducted and the blind condition is removed, i.e, the 'box' is opened. The result is very satisfactory — the event of September 14, 2015 has a high statistical significance! The statistical significance estimate is limited only by the amount of data (five days) available from time slides. If more data would have been available for the background estimation, the significance of the event would have been even higher. In conclusion, more than fifty years after Weber's first experiments, the direct detection of gravitational waves had finally been achieved! It was time to celebrate!

Another result of the analysis was the realization that there is no indication in the dataset of further, possibly weaker, signals from other binary systems that are not visible to the naked eye. Such indications could have dispelled any possible remaining doubts about the authenticity of the first and only measurement of a gravitational wave at this point in time. Once again, the researchers must practice a bit more patience.

6.1   WHAT WAS OBSERVED?

The template that best matches the observed waveform of the event on September 14th is shown in Figure 6.2. It describes gravitational waves from a compact binary system consisting of two black holes with masses of 36 and 29 times the mass of the sun. Due to the high statistical significance and the plausibility of the interpretation of the signal, it can be stated that a merging binary system of black holes have been observed! The size of the signals in both detectors indicates that the source was approximately 1.4 billion light years away from earth.

Shortly before the fusion, the black holes circled at about half the speed of light until a single black hole with about 62 solar masses was formed; this is about three solar masses less than the sum of the two black holes before the fusion. The missing three solar masses were converted into the energy of the gravitational waves, with the majority converted in only a tenth of a second. The maximum energy turnover occurred during the short merger phase, when a power equivalent of 200 solar masses per second was radiated away in gravitational waves. This is about fifty times the radiant power of electromagnetic waves (foremost visible light) emitted by all the stars in the entire observable universe at the same time!

The extremes that are in play in an observation such as this are difficult to illustrate. The astonishment of this measurement may be summarized as follows: During an event in an unimaginably distant part of the universe, a gigantic amount of energy was released and sent on its way as a gravitational wave. As it passed through the earth, 1.4 billion years later, in a fraction of a second this wave caused a tiny change in the length of an apparatus that was the product of decades of work by hundreds of scientists and engineers.

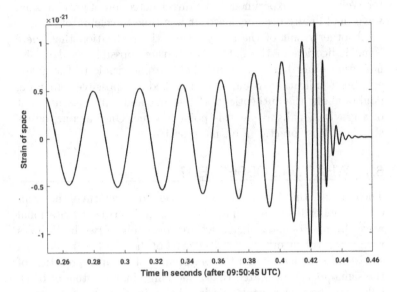

Figure 6.2 The template that best matches the observed signal shape describes the modeled stretching and shrinking of space. At earlier times, on the left side of the graph, one can see larger wave amplitudes than in the data recorded by the detectors shown in Figure 6.1. This is due to the fact that these somewhat slower oscillations are not registered very well and therefore disappear in the noise of the detector.

6.2 THE PUBLICATION

Finally, a committee is convened within the collaboration to prepare several draft articles to be published simultaneously. The process unfolds in several steps, over several weeks, during which time all members of the collaboration have the opportunity to make revisions. The title of the main publication is discussed intensively. For example, whether or not the word *direct* should be used to underscore the fact that this was a direct, rather than an indirect, observation of gravitational waves is a point discussed at particular length. This was due to an earlier discovery that pointed (disputably more indirect) to the existence of gravitational waves. Since 1975, astronomers have been measuring the orbital parameters of a particular binary system of two neutron stars using radio telescopes. This was possible because one of these neutron stars is a pulsar (see Chapter 7). It was found that this binary system, as time goes by, loses energy. This energy loss allowed researchers to conclude that gravitational waves were indeed generated, since the energy loss precisely matched the amount predicted by the General Theory of Relativity. This was regarded as indirect proof of gravitational waves.

Most scientists would most likely consider a measurement with an instrument that detects fluctuations of the space-time distortion as a direct measurement of gravitational waves, although this interpretation is also disputable. In some sense, the measurement could also be classified as indirect because complex measurements always involve several steps and the criteria are ultimately unclear. This quickly brings us into the philosophical realm when one might ask, for example, whether the human capacity for hearing or seeing can be considered 'direct', or not.

The title of the main publication is finally chosen to be *Observation of Gravitational Waves from a Binary Black Hole Merger*. In describing the observation, characteristics such as direct or first are omitted in order to keep the title simple and to avoid any possible misinterpretations with respect to different parts of the title. However, the attribute *direct* is used in the main text of the article. Many aspects of the publication were discussed in order to reach a compromise among approximately one thousand collaboration members, which is quite a feat.

As the discussions continue, more data is being collected from the detectors (16 days of observation time, in total), which allows for a more accurate estimation of the background. However, the question lingers as to whether or not the characteristics of the detector data were sufficiently homogeneous over this longer period of time to enable a valid background estimation. Nevertheless, given the additional background data collected, a consensus is reached and the event is considered to have a statistical significance of more than 5 sigma, which means that it also meets the standards of particle physics to be considered a discovery. The conclusion is that when taking simultaneous, random, disturbing events in the detectors into account, an event of this kind would be expected to occur by chance less than once in 200,000 years. The significance can only be estimated to be *more* than 5 sigma, because more data would have been required for a more accurate determination using the time-slide method.

On December 21, 2015, the tenth draft of the main publication is voted on in a collaboration-wide teleconference. It is approved in a final vote of 587 to 5. The manuscript is submitted to the journal *Physical Review Letters*, and sent by the journal to three anonymous reviewers. After minor changes, it is published on February 11, 2016, and the historic first measurement of gravitational waves is simultaneously announced in a press conferences in several involved countries. In actuality, there were three discoveries to announce: the first direct measurement of gravitational waves, the first observation of a binary system of black holes, and the first observation of the fusion of two black holes.

Within 24 hours, the manuscript is downloaded approximately 230,000 times — the most downloaded manuscript in the history of the *Physical Review Letters*. The media response is also enormous. Many daily newspapers chose the proof of gravitational waves to be the cover story. President Barack Obama congratulated the collaboration, describing this momentous achievement as a huge breakthrough in the way we understand the universe. The British astrophysicist Stephen Hawking commented: *It is a result that is at least as important as the discovery of the Higgs Boson.*[1] *Gravitational waves provide a completely new way of looking at the universe. The ability to detect them has the potential to revolutionize astronomy.*[2]

6.2.1 Further remarks

The resonant antennas, Auriga and Nautilus, were still in operation in 2015, but even if they had been as sensitive as LIGO at their resonant frequency, they would not have been able to measure the September 14th event, now officially called GW150914. As previously explained, resonant antennas are only sensitive in a narrow frequency range, a range that happens to exceed the highest frequencies that occurred during this event. In addition to the fact that interferometers are sensitive in a wide frequency band, which allows for the detection of the different phases of a merging binary system, it is possible to gain significant astrophysical insights because the waveform can be matched to a template.

At the time of the event, the GEO 600 detector in Germany was intended to be in its normal measurement mode, but happened to

---

[1] https://www.independent.co.uk/news/science/stephen-hawking-congratulates-gravitational-waves-discovery-and-says-we-can-expect-many-more-a6868151.html

[2] https://www.bbc.co.uk/news/av/science-environment-35551144/prof-stephen-hawking-celebrates-gravitational-wave-discovery

not be in the locked state, which means that no measurement data was recorded. Due to its lower sensitivity in comparison to the larger detectors, it would be an unusual stroke of luck if it were possible to measure gravitational waves with GEO 600: Astrophysical events in the near universe that would generate sufficiently large waves are extremely rare. In 2015, the Virgo detector was still in the process of being upgraded to Advanced Virgo; therefore, it also did not record any measurement data during this event. Unlike the GEO Collaboration, the members of the Virgo Collaboration are not part of the LIGO Scientific Collaboration (LSC), but a data exchange agreement exists between LIGO and Virgo, whereby both groups secure authorship rights for one another's publications. Therefore, the Virgo members were also authors of the GW150914 publication.

Because only the two LIGO detectors noted the signal, the determination of the direction from which the wave originated was not very precise, covering an area of about 600 square degrees. This corresponds to approximately three thousand times the area of the full moon as it appears to us, viewed from earth. More gravitational wave detectors with greater sensitivity would result in significant improvements in localization.

The British sociologist Harry Collins, who has followed gravitational wave research for more than forty years, freely admits, with somewhat of a guilty conscience, that a marginal first detection would have been more interesting in the realm of scientific sociology, because it would have naturally triggered more debate among experts about the interpretation of the observation. However, from the point of view of the scientists involved, the strength of the detected signal was rather a stroke of luck that ushered in the era of gravitational wave astronomy and ensured that the first direct measurement of gravitational waves was not only virtually universally accepted by experts, it was also received with great enthusiasm. On February 11, 2016, although a small amount of information had leaked out, the collaboration of more than a thousand people managed to surprise the world with the big news.

The rigour of data analysis, including the blinding technique, demonstrated in this historically first direct measurement of a gravitational wave is certainly also due to the history of the research field. In addition to the claims by Joseph Weber and the BICEP-2 collaboration (which announced the possible discovery of gravitational waves from the Big Bang in 2014), other scientists working with a cryogenic resonance antenna had previously announced the possibility of evidence for gravitational waves in their data. However, none of these groups was able to convince the experts of the interpretation of their data. With the establishment of gravitational waves as carriers of astronomical information, the question of reliable proof of their existence will no longer be an issue in the future.

6.3 OBSERVATIONS FROM THE O1 AND O2 DATA RUNS

On October 12, 2015, just four weeks after the historic first measurement of gravitational waves, another event was detected, but with a much weaker signal, achieving a statistical significance of only about 2 sigma. Although it could not yet be classified as a sure detection, the cause of the signal was most likely astrophysical, probably a binary system of black holes, once again. Soon after, on December 26, 2015, another event was registered, this time with a high significance. The signal form corresponded to a binary system of black holes with solar masses of 15 and 7 — these were significantly smaller than those of the historical first detection. Since it was not visible to the naked eye in the detector data, this signal could only be found by using templates. The event achieved a high statistical significance due to the fact that it remained in the sensitive frequency band of the detectors much longer than the first event. This was due to the lower masses of the black holes: Lower mass black holes merge at a higher frequency, creating more cycles of the inspiral phase (when the two black holes spiral towards each other) in the data. With two firm detections, any last doubts that might have remained after a single detection were eliminated for the scientists in the collaboration. With these new measurements, it was clear that the era of gravitational wave astronomy had indeed begun!

After the end of the O1 data run, the two LIGO detectors were released for commissioning in January 2016, in order to improve their sensitivity for the next data run. Interventions were made to increase the stability of the detectors, extend the time periods during which detectors are in operation and improve the quality of the acquired data. At this time, while new ground was still being broken, the limiting noise sources at low frequencies had yet to be identified. In spite of this difficulty, the second data run (O2) started on November 30, 2016 with slightly improved sensitivity. O2 continued, with two interruptions, until August 25, 2017. Due to the extraordinary efforts to complete the Advanced Virgo detector in Italy (see Chapter 4), this detector was able to participate in the O2 data run from August 1, 2017.

During this second data run of the advanced detector era, several more binary systems of black holes were registered with high significance. Particularly noteworthy was the event of August 14, 2017, when a gravitational wave was observed together with the Virgo detector for the first time. Among other things, a measurement from three detectors allowed for improved precision in the localization of the source in the sky. Table 6.1 summarizes the binary systems of black holes observed in O1 and O2.

Table 6.1 Binary systems of black holes observed in O1 and O2. The naming convention for the events consists of the letter abbreviation 'GW' and the date of detection in the form of year, month and day (YYMMDD). SNR denotes the signal-to-noise ratio of the event in the network of detectors. The estimated distance to the event is given in Megaparsecs (1 Megaparsec is approximately 3 lightyears). Estimations of the masses of the two black holes for each event are given in units of the solar mass $M_\odot$.

Name	SNR	Distance [Mpc]	$M_1/M_\odot$	$M_2/M_\odot$
GW150914	24.4	430^{+150}_{-170}	$35.6^{+4.8}_{-3.8}$	$30.6^{+3.0}_{-4.4}$
GW151012	10.0	1060^{+540}_{-480}	$23.3^{+14.0}_{-5.5}$	$13.6^{+4.1}_{-4.8}$
GW151226	13.1	440^{+180}_{-190}	$13.7^{+8.8}_{-3.2}$	$7.7^{+2.2}_{-2.6}$
GW170104	13.0	960^{+430}_{-410}	$31.0^{+7.2}_{-5.6}$	$20.1^{+4.9}_{-4.5}$
GW170608	14.9	320^{+120}_{-110}	$10.9^{+5.3}_{-1.7}$	$7.6^{+1.3}_{-2.1}$
GW170729	10.8	2750^{+1350}_{-1320}	$50.6^{+16.6}_{-10.2}$	$34.3^{+9.1}_{-10.1}$
GW170809	12.4	990^{+320}_{-380}	$35.2^{+8.3}_{-6.0}$	$23.8^{+5.2}_{-5.1}$
GW170814	15.9	580^{+160}_{-210}	$30.7^{+5.7}_{-3.0}$	$25.3^{+2.9}_{-4.1}$
GW170818	11.3	1020^{+430}_{-360}	$35.5^{+7.5}_{-4.7}$	$26.8^{+4.3}_{-5.2}$
GW170823	11.5	1850^{+840}_{-840}	$39.6^{+10.0}_{-6.6}$	$29.4^{+6.3}_{-7.1}$

The subscripts and superscripts in the table denote the 90 % credible intervals, which means that the true values lie within the given range with a probability of 90 %. The combined mass (the so-called chirp mass) of the two black holes in each detection is known much more accurately.

These frequent observations of gravitational waves resulting from merging black holes came somewhat as a surprise, given the fact that most collaboration researchers considered it to be more likely that the first detection would be a result of the inspiralling and merging of two neutron stars.

In addition to estimating how often binary black holes merge throughout the observable universe, there are further intriguing conclusions that can be drawn from these early observations:

- About half of the black holes from the detected binary systems have solar masses of more than 20. The formation of such massive black holes (distinct from the supermassive black holes in the center of galaxies) from the individual development of stars was previously considered unlikely. As a result of this, existing models about the evolution of stars must be newly considered and may need to be expanded, or alternative ways of how these massive black holes have formed need to be proposed.

- The waveforms calculated using the General Theory of Relativity and the actual waveforms observed are a striking match. There is no evidence of any deviation from Einstein's theory — once again, it is brilliantly confirmed. Einstein was right! This also applies to the polarization of gravitational waves, observed for the first time with the GW170814 event, which was made possible by the additional measurement of the Virgo detector.

- Finally, given the observations to date, an interesting upper limit for the dispersion of gravitational waves was determined. In some alternative theories of gravity, dispersion describes the possibility that different frequencies of gravitational waves travel at different speeds. However, the measurements of gravitational waves again show no deviation from the general theory of relativity, according to which the waves propagate at the speed of light and dispersion is zero.

How are the observed binary systems of black holes formed? This cannot yet be answered definitively, but both of the previously assumed pathways of formation (Chapter 1) are compatible with observations. These are: formation from binary systems of stars and formation in dense stellar environments through interaction with other objects.

6.3.1 Merging neutron stars!

Just as the O1 data run began with the detection of GW150914, the O2 data run ended with another bang. For the first time, on August 17, 2017, the two LIGO detectors and the Virgo detector received a gravitational wave signal from two merging neutron stars! The signal, which was observed for almost 100 seconds in the detector measuring ranges, came from a distance of only 40 Mpc (about 120 light-years) and had a signal-to-noise ratio of 32 — the clearest gravitational wave measured in all of the O1 and O2 data runs. Figure 6.3 shows the time course of the frequency of the observed signal. The total mass of the objects in this binary system was estimated to be 2.74 solar masses. This is essentially determined by comparing the signal curve with possible templates. However, since the mass of each individual object was in the range of 0.86 to

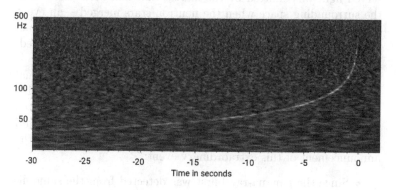

Figure 6.3 The last 30 seconds in the 'life' of two neutron stars: The event GW170817. The bright track (combined from the data of the two LIGO detectors) shows the measured gravitational wave signal (the strain of space-time at the location of the detectors) resulting from the two objects orbiting each other. Towards the right, over time, the upward curved shape shows the increasing speed of the orbiting neutron stars. Just before merging, at zero seconds, the stars move around each other almost 250 times per second, resulting in gravitational waves of nearly 500 cycles per second (500 Hz).

2.26 solar masses, one of the objects could have been a black hole, but researchers concluded it was most likely two neutron stars, based on various astronomical observations and models.

The first observation of the fusion of two neutron stars by the detection of gravitational waves was spectacular enough in itself, but there was more! Almost simultaneously (only 1.7 seconds after the fusion), the satellites Fermi and INTEGRAL received a short burst of gamma rays. An alarm was sent out to astronomers by the gravitational wave collaboration, identifying a patch of sky, almost 30 square degrees, as the direction from which the gravitational wave originated. About 11 hours after the registration of the gravitational wave event, the SWOPE telescope in Chile was the first to sight a new illuminated object in the galaxy NGC4993. Within the next few hours, this new object was detected by five other telescopes. Subsequent observations by up to seventy observatories across the world (in the radio, infrared, optical and X-ray range) followed the luminosity of the object, which was initially strong, and then become weaker. It is now considered certain that the observed light was emitted by the matter that had been hurled into the surrounding space when the neutron stars merged — an event called a *kilonova*. The magnitude of this worldwide concerted action, for which high-level observation programs were interrupted, was unprecedented in the history of astronomy.

The simultaneous observation of an event using gravitational wave detectors *and* classical telescopes introduced the era of multi-messenger astronomy with gravitational waves. Only time will tell us the full extent of the scientific richness of these observations and others of their kind, but some conclusions could be drawn at the announcement of this extraordinary event:

- Since the gamma-ray burst was detected from the same direction as the gravitational waves, this event is considered to be evidence that short gamma-ray bursts can actually be generated by merging neutron stars. Prior to this event, this process was only a suspected cause of such gamma-ray bursts. The reason for a 1.7-second delay in the gamma-ray flash could only be speculated on and further observations are necessary to gain a more precise understanding.

- By observing the gravitational wave signal and the gamma-ray burst almost simultaneously, a very narrow limit for the traveling speed of the gravitational waves could now be determined: As predicted by general relativity, they propagate at the speed of light, with a possible deviation of less than 10^{-15}, thus excluding some alternative theories of gravity that had assumed different values for the propagation speed of gravitational waves.

- For some time, supernovae were thought to be the dominant breeding grounds for heavy chemical elements, but doubts had recently arisen as to whether the supernovae theory was sufficient to fully explain the abundance of these elements in the universe. During the merging of the neutron stars of GW170817, about 0.1 to 1 % of their matter was hurled into the surrounding space. For the first time, spectroscopic observation of this matter proved that many heavy elements, such as gold, platinum, lead and uranium, are produced during such an event. It appears likely that matter in a quantity several thousand times the mass of the earth was created in the form of heavy elements during GW170718 — the quantity of gold created alone was many times the mass of earth!

- According to theory, the fusion of the two neutron stars initially led to the formation of a very massive single neutron star. Due to its high mass, it was likely to quickly collapse into a black hole, but it could have survived for a period of time, because its rapid rotation could keep it from collapsing. Nine days after the fusion event, the X-ray satellite, Chandra, discovered a flash of X-rays of the ejected matter. This may indicate that, before collapsing into a black hole, the neutron star actually survived for a few days emitting high-energy radiation. Researchers differ, however, in their interpretation of this observation.

- The shape of the gravitational waves from merging binary systems can be used to draw conclusions about the masses of the objects involved and, thus, conclusions about the strength of the gravitational waves at their source location. If the magnitudes of the signal observed in the detectors are precisely measured, the distance of the source can thus be estimated. This is because the waves attenuate proportionally to the distance of their journey through the universe. In the case of the observed neutron stars, the corresponding galaxy could be determined (NGC4993), which allowed for a measurement of its velocity (via the redshift of the observed light). With this data, from gravitational-wave and optical observations, it is possible to measure the velocity of the expansion of the universe as a function of distance (the Hubble constant), a method that had been proposed by Bernard Schutz of Cardiff University. The Hubble constant that was determined with these measurements is consistent with the previously known values. With this method it can be determined more directly, however, and future measurements will provide more accurate results.

It is fascinating to realize that the heavy elements abundant in our universe originated, to a large extent, from merging neutron stars. Heavy elements are an important ingredient in the formation and dynamics of planets and, therefore, ultimately for all life forms known to humankind. The reader may be reminded that heavy radioactive elements within the earth are the main heat source that drives plate tectonics, being an essential ingedient in the evolution of life. Without the existence of gravitational waves and their radiation, the neutron stars would not have merged (see Chapter 1 and the two-body problem). In addition to many other physical laws and properties, gravitational waves are also responsible and are required for the existence of the world as we know it!

Finally, while these results are spectacular, it should be mentioned that gravitational waves in other forms, i.e., impulsive, quasi-continuous, or as stochastic noise, have not yet been found in the data as of summer 2019. Curiosity continues to fuel discovery!

6.4   BEYOND O2

After the end of the O2 observing run, the LIGO and Virgo detectors were once more handed over to the commissioning teams, in order to further increase their astrophysical reach. Several improvements were implemented over the course of almost 2 years, and we give examples of some of these:

- Both LIGO and Virgo implemented the squeezed vacuum technique that had been in operation at the GEO 600 detector since 2011.

- Some more free space was created around the main test masses of LIGO to reduce the number of impacts from residual gas molecules hitting them.

- Some of the coatings of the mirrors that make them highly reflective were improved to reduce the amount of light being scattered.

- More baffles were installed in order to further reduce stray light that otherwise disturbs the measurement.

- Dampers were installed on the LIGO test masses to reduce oscillations that can occur when operating with high laser power (so-called parametric instabilities that stem from the interaction of the laser light with the vibrations of the test masses).

With these and other improvements, the O3 science run started on April 1st, 2019 with further impoved sensitivity. The LIGO detectors have now reached a range of 100 to 140 Mpc for binary neutron star inspiral signals, and the Virgo detector has reached between 40 and 50 Mpc. During the O3 run, the network sends out triggers of potential signals shortly after detection to alert the astronomy community. After less than a month into the run, already three more binary black hole merger events had been detected, and on April 25, 2019 another binary neutron star merger was very likely registered. On August 14, 2019 the merger event of a black hole and a neutron star was likely identified, the first ever event of this kind. Gravitational wave detections now happen around once per week on average, confirming the expected rapid growth of signal numbers as the detectors get more sensitive.

The O3 run is foreseen to last one year, and will be followed by another commissioning break. Observing runs O4 and O5, with successively higher sensitivity and more frequent detections, are foreseen for 2021 to 2022, and from late 2023 onwards, respectively.

CHAPTER 7

Future developments

After 50 years of effort, large-scale physics experiments, the gravitational wave detectors, confirmed the existence of gravitational waves with the spectacular detections of coalescing black holes and neutron stars. At the same time, these events demonstrated the profound usefulness of gravitational wave detectors as a new tool for making astronomical observations. From the very beginning, the observational aspect was a strong motivational force in the development of this branch of research. Using knowledge gleaned from the detection of more gravitational waves, scientists expect that a series of astrophysical, cosmological, and fundamental-physics puzzles will eventually be solved. These questions include: What is the matter of neutron stars like? How does a supernova explosion occur? How are black holes and compact binary systems created and how do they develop? What does the cosmic distribution and merging rate of compact binary systems tell us about the evolution of the universe? How fast does the universe expand? Does Einstein's General Theory of Relativity correctly describe gravity near black holes or must alternate theories of gravity be considered? What happened right after the Big Bang?

If humanity chooses to attempt to find answers to these and other questions, it is necessary to improve the performance of existing detectors or to develop new interferometers on earth. In addition, the accessible frequency range of gravitational waves can be expanded by placing detectors in space (see "LISA" below).

7.1 EARTHBOUND INTERFEROMETERS

In the coming years (after 2019), the goal is to achieve the planned-for sensitivity in Advanced LIGO, Advanced Virgo and KAGRA and, eventually, LIGO-India. In order to increase the earth-bound detectors' sensitivities beyond their initially targeted astrophysical range, three factors (discussed in Chaper 3) are key: longer arms, quieter mirrors and more light.

While longer arms can only be achieved with new infrastructure, quieter mirrors can be achieved with further development of seismic isolation systems and optics. More light can be provided for detection with the development of stronger lasers; however, due to thermal effects, it will most likely be necessary to use alternate materials for the mirrors and different laser wavelengths to be compatible with those materials. This is an active branch of research. As a possible path forward, using silicon as the material for the mirrors and doubling the wavelength of laser light to about 2 micrometers are currently being investigated.

In the near future, however, effects in the interferometers that were previously of less relevance will have to be taken into account and overcome. These include: radiation pressure noise, Newtonian noise, and thermal noise.

Radiation pressure noise

To improve the sensitivity of the detectors at higher frequencies, the amount of light in the arms must be increased. If the light power in the interferometer exceeds a particular level, radiation pressure noise will begin to reduce sensitivity at lower frequencies. Similar to shot noise, radiation pressure noise results from the quantum interaction of light with matter. Each light photon reflected from a test mass transmits a small mechanical impulse, triggering a tiny movement of the mass. The multitude of photons in a light beam produces a characteristic noise in the position of the test masses, the radiation pressure noise. One way to counteract the problem is by making the test masses heavier. In this way, the movements caused by the photons become smaller, since the heavier test masses resist acceleration from the photons more. However, there are also practical limits to the size of a test mass and radiation pressure noise can still limit gravitational wave measurement.

If the radiation pressure noise in the detector limits sensitivity, the application of squeezed vacuum becomes more difficult. While squeezed vacuum reduces shot noise (see Chapter 4), it increases radiation pressure noise; this effect can be explained by the vacuum fluctuations. In the future, to reduce this undesirable effect, optical resonators will be required to filter the squeezed vacuum before it can be suitably applied to the interferometer. The filtered squeezed vacuum will allow to reduce shot noise and radiation pressure noise at the same time.

Newtonian noise

To date, Newtonian noise has been masked by other, bigger noise sources. In the future, it will become an issue at low frequencies of the detection spectrum for ground-based detectors. As the name indicates, Newtonian noise relates to Newton's theory of gravitation. As discussed in Chapter 1, his theory describes all masses as attracting each other. Although Einstein's theory of gravity is more accurate than Newton's with regard to massive celestial objects at close ranges, Newton's theory and its corresponding descriptions are still relevant and continue to be applied in both physics and engineering. It can be said, then, that any form of mass in the vicinity of the test masses in the interferometer exerts a small attractive force directly upon the test masses, essentially bypassing the seismic isolation. If these forces did not change, they would not disturb the measuring process. However, when masses move in the environment, the Newtonian force they exert on the test masses also changes — an effect that can hardly be distinguished from that of a gravitational wave.

The main masses that can move in the close environment are particles that compose the soil. Movements caused by natural and man-made vibrations lead to density fluctuations of the soil and the air, and thus to fluctuating Newtonian forces on the test masses. As a countermeasure, to reduce these vibrations, ditches can be used or future interferometers can be built underground, as already happened for the KAGRA interferometer in Japan. In addition, there is the possible use of a variety and large quantity of seismometers to detect ground movements in the vicinity of the test masses. The information gathered can then be utilized to calculate the expected Newtonian noise, which can then be subtracted, to some extent, from the output signal of the interferometer.

Thermal noise

For interferometer components that are kept at room temperature, the most severe source of thermal noise comes from the optical coatings of the test masses. The coatings are required to let the test masses function as mirrors with extreme high reflectivities; however, the mechanical friction losses within the coating materials is a limiting source of noise. Intensive research into new materials and material composition is going on, with the goal of reducing the mechanical losses, and thus the coating thermal noise.

Another way forward to counteract thermal noise is operating at low temperatures, a path that was also taken in the development of resonant mass antennas (see Chapter 2). In addition to being located underground, the KAGRA project is also doing pioneering work on reducing thermal noise by cooling the test masses and their suspensions. Just as in the case of cylinder antennas, however, cooling does increase the complexity of the detectors even more. For example, cooling must be done in a way that does not compromise the seismic isolation of the test masses. This means that parts of the cooling systems themselves must be seismically isolated in a similar way as the test masses are. In addition, the cooling system must be designed to tolerate the constant heat input to the test masses resulting from a small fraction of laser light that is inevitably absorbed.

Possible new sources of gravitational waves at the low-frequency end of earth-based detectors are more massive binary black holes that merge at a lower frequency. On the high-frequency end, around 2 kHz, more sensitivity will make it possible to measure fine details of the gravitational waves immediately after two neutron stars merged. This type of measurement will allow more detailed insight into the composition of the very dense neutron star matter.

In addition to the gradual improvement of existing interferometers and the integration of new generations of these instruments into existing infrastructures, however, there are also plans for completely new facilities on earth.

7.1.1 The Einstein Telescope

The Einstein Telescope (ET) is the European concept of a new gravitational-wave detector facility. From 2007 to 2011, a design study for ET was prepared, and updates have been gradual and ongoing. To reduce Newtonian noise and seismic vibrations, the plan is to build ET underground at a depth of at least one hundred meters (328 feet), where seismic noise is already significantly reduced. ET's infrastructure, which is planned for a service life of at least fifty years, may consist of three tunnel tubes, each ten kilometers long (6.2 miles), arranged in a triangle with several caverns dug up at the corner points. Figure 7.1 shows an artist's conception of ET.

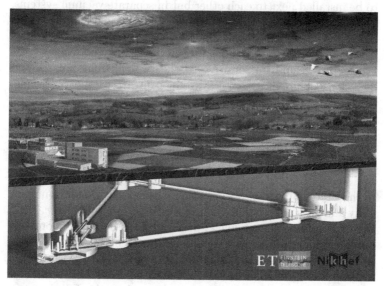

Figure 7.1 An artistic view of the Einstein Telescope. (Image courtesy of Nikhef.)

A new prospect is emerging in the design of ET whereby individual interferometers are optimized specifically for a certain frequency range of gravitational waves. In some sense, this resembles the historic development of optical telescopes, which, over time, became more and more specialized in order to detect particular wavelengths of light and other electromagnetic radiation. This principle is also likely to be followed in the case of gravitational wave detectors, particularly because the techniques for increasing sensitivity at low and high frequencies are different and are not compatible. For example, in order to improve sensitivity at low frequencies, the test masses and their suspensions need to be cooled, while to improve sensitivity at high frequencies, more laser light is needed in the interferometer; if more light is used, the heat input to the test masses increases, which makes cooling more difficult. In addition, as light power increases, radiation pressure noise increases (as mentioned above) and this can also impair sensitivity at low frequencies. As of 2019, two interferometer types are planned for ET: one with very high light power in the arms (3 megawatts) mostly sensitive for frequencies above about 30 Hz, and one with lower light power and cryogenic test masses optimized for operation at lower frequencies. Both interferometer types are to be installed next to each other but in separate vacuum systems. This concept of constructing dedicated interferometers for specific frequency bands is also called a *xylophone* configuration.

If ET is built with three tunnels in a triangular shape, the ET facility would be sensitive to the two different polarization directions of gravitational waves. In contrast to the traditional shape with perpendicular interferometer arms, this would require the construction of a much more complex facility with more interferometers.

7.1.2 Cosmic Explorer

In the United States, a concept study is underway to build a new facility and detector as well, a concept called *Cosmic Explorer*. Here, the plan is to have even longer arms, perhaps up to 40 kilometers long, but construct the interferometer on the surface of the earth, to save cost in comparison to an underground facility. If the cosmic explorer facility is indeed built on the surface, one has to sacrifice a bit of low-frequency sensitivity compared to the ET detector, mostly due to Newtonian noise. In turn, the Cosmic Explorer detector can be more sensitive than ET at higher frequencies, if its interferometer arms will be longer. But since it will likely be a traditional L-shape configuration with a single detector, it will not be sensitive to both polarizations of gravitational waves and will rely on other detectors in a network to retrieve this information.

For such long interferometer arms as foreseen for Cosmic Explorer, some new challenges have to be overcome. One stems from the fact that the suspended mirrors hang vertically with respect to the center of mass of the earth at each of the endpoints of the long arms, such that the suspension chains will not hang precisely parallel. This introduces an increased coupling of vertical seismic noise to the mirror position, and thus demands an improvement in vertical isolation technology. Another challenge with very long arms comes from the large beam spots that are required for laser beams at these distances. Even bigger mirrors will be required while keeping the precision and optical coating quality needed for this kind of interferometry.

7.1.3 Speed measurement?

Further development of low-frequency ground-based gravitational wave detectors could also lead to a departure from the principle of the Michelson interferometer. In a Michelson interferometer, the position of the test masses is measured very precisely, but according to Heisenberg's uncertainty principle, the location (position) and the speed (or momentum, the product of speed and mass) of an object cannot be measured simultaneously with perfect accuracy. This means that gravitational wave detectors of the Michelson type will eventually be so sensitive that the extremely precise position measurement of the test masses will lead to uncertainty about the speed of the mirrors. This uncertainty has a disturbing effect on the position of the test mass — its exact location cannot be determined the next time the location is measured, i.e., a short time later. Since one wants to measure the location continuously, there will always be a disturbance from the measurement that happened just a moment ago. (This effect is equivalent to radiation pressure noise, as it turns out.) Although it may sound a bit magical, scientists can work around this problem by directly measuring the speed of the test masses instead of the position. This might introduce greater inaccuracy in the determination of the position of the test mass during the next speed measurement, but it's not a problem since position is no longer a relevant measurement variable in this scenario. This kind of speed measurement, also called a *speed meter*, may be realized in the future with optical configurations such as Sagnac interferometers.

7.2 SEARCHES AT OTHER FREQUENCIES OF THE SPECTRUM

This history of the prediction and eventual measurement of gravitational waves has focused on the use of ground-based detectors, resonant antennas and interferometers. Ground-based interferometers are currently sensitive in the range of roughly 10 to 10,000 Hz, which is sometimes described as the high-frequency portion of the gravitational wave spectrum. Currently, two promising technologies for measuring gravitational waves at frequencies lower than those accessible with earth-bound detectors are under development — space interferometers and pulsar timing. Each represents a complement to earth bound detectors in that they extend the detectable frequency spectrum. This is comparable to the development of radio telescopes which have extended the detectable frequency spectrum of optical and electromagnetic astronomy.

7.2.1 LISA

The most prominent project by far in space interferometry is LISA (Laser Interferometer Space Antenna). The LISA project is supported by the European Space Agency (ESA) and the LISA consortium, which consists of twelve European member countries and the United States. The LISA project, led by Karsten Danzmann, was selected by ESA to be an L3 mission, which means that LISA is scheduled to be launched into space in 2034.

The LISA mission consists of three identical satellites which follow the earth around the sun at a distance of about fifty million kilometers (over 31 million miles). The satellites will be in a triangular configuration at a distance of about 2.5 million kilometers from one another (1.5 million miles). They will send each other laser beams in order to measure small travel time differences caused by gravitational waves that squeeze and stretch space-time between the satellites. Figure 7.2 shows a cartoon of the three LISA satellites in their orbit around the sun.

The reference points for the laser beams are two, free-floating, cubes located inside the satellites, with an edge length of 46 millimeters (1.8 inches) made of a non-magnetic gold-platinum alloy. Each satellite automatically positions itself around these almost free-floating test masses using control loops. This technology, called drag-free control, was tested very successfully in 2016 with the test satellite LISA Test Package (LTP). This mission flew two test masses within one satellite and measured their distance to picometer precision, while both cubes were floating (nearly) freely.

Space technology offers two advantages over interferometers on earth. First, the interferometer is less disturbed by environmental influences that can limit measurement on earth — there are virtually no seismic shocks, and Newtonian noise caused by thermal expansion of the satellites is minimal. Second, the arms of the interferometer, i.e., the distance between the satellites, can be much longer than those on earth due to the almost perfect vacuum in space and the fact that earth is simply too small for such large interferometers — measuring distances on earth are approximately

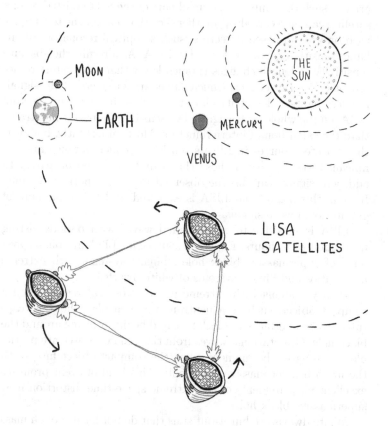

Figure 7.2 A cartoon of the three LISA satellites in their orbit around the sun. The drawing is not to scale! (Illustration: Josh Field.)

one million times shorter! Longer arms allow more signal (i.e., phase shift of the laser light by a gravitational wave) to be collected. However, very long arms have a disadvantage as well: Since the laser beams expand in size when travelling from one satellite to another, the light beam is much larger than the satellite where it arrives, such that only a very small part of the light emitted by one satellite can be received by another. For this reason, the techniques used on ground-based detectors, such as optical resonators to enhance sensitivity, cannot be used in LISA. As a result, the absolute sensitivity to length changes is much lower than with interferometers on earth. Both effects (much longer arms to increase the signal, but less light to detect it) almost cancel each other out, making LISA, on the whole, only slightly less sensitive to space-time strains than the earth-bound interferometers. Most importantly, however, the expected sources in LISA's sensitive frequency range are significantly stronger than those that can be received on earth. In addition, signals can also be observed for longer periods of time. In sum, this means that LISA is expected to detect a plethora of gravitational wave signals.

LISA is sensitive to gravitational waves with periods lasting from minutes to hours. Expected sources for LISA are binary systems of supermassive black holes, binary systems with extreme mass ratios and binary systems of white dwarfs.

Binary systems with extreme mass ratios are created when a compact object, such as a neutron star or a black hole, is 'captured' by a supermassive black hole and begins orbiting around the black hole. Gravitational waves from this kind of source are particularly strong at the moment when the compact object fuses with the much heavier massive black hole. This kind of event promises excellent measurements of the extreme space-time distortion near supermassive black holes.

White dwarfs are burnt-out suns that do not have enough mass to be condensed into a neutron star or black hole. There is a large abundance of these corpses of dead suns throughout the universe and many of them form binary systems, having originated from binary systems of regular stars. Many individual binary systems of white dwarfs can be detected by LISA, but due to their large quantity, not all of these signals may be resolved as individual sources. If many of these sources are unresolved, they will form a stochastic background of gravitational waves, as mentioned in Chapter 5. Since the LISA mission consists of three interferometers, a stochastic gravitational-wave noise can be detected by correlating the data of the individual interferometers against each other.

It is interesting to note that LISA will be able to observe binary systems for years to weeks before they merge. With LISA observations, it is expected to be possible to predict, for some of the binary systems, when they will merge in the future in the frequency band of the ground-based detectors like LIGO, Virgo, and KAGRA, or ET and Cosmic Explorer. These predictions can reach an accuracy for the time of the merger event of better than 10 seconds, and a prediction of the position of the source in the sky of less than one square degree. Had LISA been operating years before the GW150914 event detected by LIGO, the time and sky position could have been predicted by LISA! In the future, such information will allow optical (or radio and X-ray) telescopes on earth to point to the predicted patch in the sky at the expected time of the merger event, such that the fusion of black holes can be observed 'live' with gravitational waves and electromagnetic waves.

7.2.2   Other laser-interferometer projects in space

While LISA is the most mature and firmly established project, there are several other ideas at different levels of development that range from concept studies to more serious and ongoing space interferometry projects.

One class of such projects would utilize the idea of having three satellites in orbit not around the sun, but around the earth. Such orbits are easier to reach and thus the cost of the mission can be lower. At the same time, care has to be taken to sufficiently shield the test masses in the satellites from influences from the earth, such as varying gravitational forces or electromagnetic radiation.

A Japanese project called DECIGO is in a conceptual planning stage and would be sensitive in the range from tens of millihertz to a few Hertz, which would bridge the gap in the frequency band between LISA and the earth-based detectors. The satellites of DECIGO would have a distance of about 1000 kilometers between each pair of spacecraft and, in contrast to LISA, could use Fabry-Perot resonators in the optical paths between the satellites.

The People's Republic of China has embarked on a project called Tianqin, meaning *harp in the sky*, also consisting of three satellites in an orbit around the earth. The distance between each pair of spacecraft in this case would be about 170,000 kilometers, about 15 times smaller than in LISA, and thus the detector would be sensitive for a slightly higher frequency band than LISA, but a bit below the frequency band of DECIGO. The launch of a test satellite for the mission is foreseen for 2019.

Another class of laser-interferometric space detector would be a mission of at least three satellites in an orbit around the sun, but where the three satellites are even much further apart than in the case of LISA. ASTROD-GW is such a concept idea with a distance between the spacecrafts of 260 million kilometers, to be sensitive to gravitational waves below the LISA frequency band. There are plenty of ideas for other projects, and as this field of research evolves, surely new ones will emerge.

7.2.3 Pulsar timing

Since most galaxies probably have a supermassive black hole in their center, the collision of two galaxies can lead to the formation of a binary system composed of their two central black holes, which will then slowly start to orbit each other. Due to the large mass of these black holes, this type of source would make very long gravitational waves, which may in some cases be detected by LISA, but it is pulsar timing technology that is better suited to detect these very long waves.

Pulsar timing is a slightly different technology for measuring gravitational waves than that of LISA and the earth-bound interferometers. While the latter use lasers to measure space-time distortions, pulsar timing uses electromagnetic signals from pulsars received by radio telescopes on earth for that purpose.

As mentioned before, rotating neutron stars are considered to be a source of quasi-continuous gravitational waves. With pulsar timing, it is not the gravitational waves emitted by pulsars that are of interest, but the gravitational waves from other sources that pass the space between a rotating neutron star and earth. Pulsar timing makes use of the fact that neutron stars can also emit electromagnetic radiation, which is produced as they rotate around their own axis. In some cases, this rotation leads to pulses of electromagnetic radiation that can be received on earth, which is why these neutron stars are called pulsars. In 1967, by chance, Jocelyn Bell discovered the first pulsar in the data of a radio telescope. Due to the extremely large regularity of the pulses, whose timing can be as accurate as the pulse rate of an atomic clock, it was initially considered possible that the signals originated from extraterrestrial civilizations! A short time later, the generation of pulses by a rotating neutron star became a more convincing explanation.

As of today, more than 2500 pulsars have been identified — almost all of them in the Milky Way. The pulsar timing method focuses on pulsars that emit particularly clear signals which are well known and characterized. These continuous streams of electromagnetic pulses pass through the interstellar space of the Milky Way which is, among other things, filled with very long-period gravitational waves. These gravitational waves change the propagation times of the pulsar signals (by a tiny amount) on their way to earth. This means that having passed through space distorted by gravitational waves, the pulsar signals arrive at the radio telescopes on earth at a slightly different time than they would have if no gravitational wave was present on their journey. In order to detect a gravitational wave with this method, the pulsars must be observed over long periods of time so that patterns in the change of the arrival time of the pulse signals can be identified. In summary, interferometers monitor the differences in the propagation time of laser light to deduce gravitational waves, while pulsar timing monitors the arrival time of electromagnetic signals from pulsars in the Milky Way to this end.

The search for such signals is primarily carried out by three so-called pulsar timing arrays using radio telescopes in Europe, the United States, and Australia: The European Pulsar Timing Array (EPTA), the North American Nanohertz Observatory for gravitational waves (NANOGrav), and the Parks Pulsar Timing Array (PPTA) in Australia. The total number of observed pulsars is around 50, with some of them being observed by several radio telescopes. The three teams also combine their data in an international consortium, the International Pulsar Timing Array (IPTA), to obtain the best possible sensitivity using all available data.

The astronomical objects that generate gravitational waves that are expected to be measured with the pulsar timing method are binary systems of supermassive black holes that orbit each other very slowly (with periods from months to a hundred years). A stochastic background of many such signals is likely the first gravitational-wave signature to be found by the pulsar timing arrays, but an individual system of two supermassive black holes may be found as well. Other, even more exotic, very slow gravitational waves may stem from so-called cosmic strings or superstrings, which may have formed in the early universe. As of autumn 2019, no gravitational waves have been detected by pulsar timing arrays, but the results can already rule out some galaxy evolution scenarios. There is every reason to believe a detection will happen in the not too distant future as more data is acquired, the detection technology progresses, and more suitable pulsars may be discovered that can be included in the analysis.

While gravitational waves from pairs of supermassive black holes have not yet been observed, at least something close to an 'image' of an individual supermassive black hole has recently been obtained from electromagnetic observations. Figure 7.3 shows the shadow of a black hole obtained by the Event Horizon Telescope, a collaboration that used several radio telescopes in a worldwide network to produce the image.

Figure 7.3 This image, constructed from radio-telescope data, shows the shadow of a black hole in the center of the bright ring of matter. This is the central supermassive black hole in the Galaxy M87, with a mass of about 2.4 billion times the mass of the sun. (Image credit: Event Horizon Telescope Collaboration.)

7.2.4 The very high frequency end

The search for gravitational waves at very low frequencies is expected to lead to exciting new discoveries. There is no principle reason though why gravitational waves could not also exist at very high frequencies, in the range from megahertz to terahertz and beyond. (A terahertz is 10^{12} oscillations per second, i.e., 1 million megahertz.) At the end of this outlook, it seems worthwhile to mention that some experiments have been on the hunt for gravitational waves at such high frequencies. Nothing has been found, but the search seems justified, since a few possible sources of high-frequency gravitational waves may exist. So-called primordial black holes, produced shortly after the Big Bang, may have evaporated due to Hawking radiation, and in that process may have produced another stochastic background of gravitational waves. If it ever can be found, it would be yet another milestone for gravitational-wave physics and for new knowledge about the cosmos we live in.

Literature

The literature that has been consulted for this book and other literature recommended for further reading are referenced in the Bibliography.

A broad view on the development of Einstein's General Theory of Relativity can be found in the book of Ferreira [9]. A concise history of the prediction of gravitational waves from the theory of general relativity is given in Kennefick's book [12]. Coles gives a brief description of some of the history around the 1919 solar eclipse in [4], and a brief analysis of the role of the Michelson-Morley experiment in the development of Special Relativity is given by van Dongen [15].

A concise historic overview of resonant mass detectors can be found in the article from Aguiar [2], and more in-depth in the book *Gravity's Shadow* [5] by sociologist Harry Collins. The latter and two other books of Collins give an in-depth sociological account of the field of gravitational wave detection, and although not explicitly written as such, also serve as a resource of historical material: *Gravity's Shadow* [5], *Gravity's Ghost and the Big Dog* [6] and *Gravity's Kiss* [7]. The latter gives a detailed chronology of the events around the first detection.

A compact overview of the general history of gravitational wave detection can be found in the article by Cervantes-Cota, Galindo-Uribarri and Smoot [3]. The editor of *Physical Review Letters* that handled the publication of the first detected gravitational-wave event wrote an interesting commentary on the occasion in [10].

The 1989 proposals for the LIGO and GEO detectors are accessible online at [16] and [11], respectively.

Advanced Reading

A comprehensive recent overview of General Relativity is given in the 2-volume book [13], including very acessible summaries and introductions to sub-fields and the history of different aspects of General Relativity.

More details of the principles of gravitational-wave laser interferometry and the design and functioning of the Advanced LIGO and Virgo detectors can be found in the 2-volume book [14].

The design study for the Einstein Telescope can be found here [1] and the LISA proposal to ESA can be downloaded here [8].

Bibliography

[1] M. Abernathy et al. Einstein gravitational wave telescope conceptual design study. Technical Report ET-0106C-10, European Gravitational Observatory, 2011.

[2] Odylio Denys Aguiar. Past, present and future of the Resonant-Mass gravitational wave detectors. *Research in Astronomy and Astrophysics*, 11(1):1–42, dec 2010.

[3] Jorge L. Cervantes-Cota, Salvador Galindo-Uribarri, and George F. Smoot. A Brief History of Gravitational Waves. *Universe*, 2(3), 2016.

[4] Peter Coles. Einstein, Eddington and the 1919 Eclipse, 2001. arXiv 0102462 (https://arxiv.org/pdf/astro-ph/0102462.pdf)

[5] Harry Collins. *Gravity's Shadow*. University of Chicago Press, 2010.

[6] Harry Collins. *Gravity's Ghost and Big Dog: Scientific Discovery and Social Analysis in the Twenty-First Century*. University of Chicago Press, 2014.

[7] Harry Collins. *Gravity's Kiss: The Detection of Gravitational Waves*. MIT Press, 2017.

[8] Karsten Danzmann et al. LISA Laser Interferometer Space Antenna, A proposal in response to the ESA call for L3 mission concepts, 2017. https://www.elisascience.org/files/publications/LISA_L3_20170120.pdf accessed 9/18/2019.

[9] Pedro G. Ferreira. *The Perfect Theory: A Century of Geniuses and the Battle over General Relativity*. Houghton Mifflin Harcourt, 2014.

[10] Robert Garisto. Commentary: How gravitational waves went from a whisper to a shout. *Physics Today*, 69(8):10, 2016.

[11] J. Hough et al. Proposal for a joint German-British interferometric gravitational wave detector. Technical report, 1989.

[12] Daniel Kennefick. *Traveling at the Speed of Thought*. Princeton University Press, 2007.

[13] Wei-Tou Ni, editor. *One Hundred Years of General Relativity*. World Scientific, 2017.

[14] D. Reitze, P. Saulson, and H. Grote, editors. *Advanced Interferometric Gravitational-Wave Detectors*. World Scientific, 2019.

[15] Jeroen van Dongen. On the Role of the Michelson-Morley Experiment: Einstein in Chicago. *Arch. Hist. Exact Sci.*, 63:655–663, 2009.

[16] R.E. Vogt, R.W.P. Drever, F.J. Raab, K.S. Thorne, and R. Weiss. The Construction, Operation, and Supporting Research and Development of a Laser Interferometer Gravitational - Wave Observatory, Proposal to the National Science Foundation submitted by the California Institute of Technology, unpublished; available online at https:// dcc.ligo.org/public/0065/M890001/003/M890001-03 Technical report, 1989.

Index